DESTINATION
MARS

IN ART, MYTH, AND SCIENCE

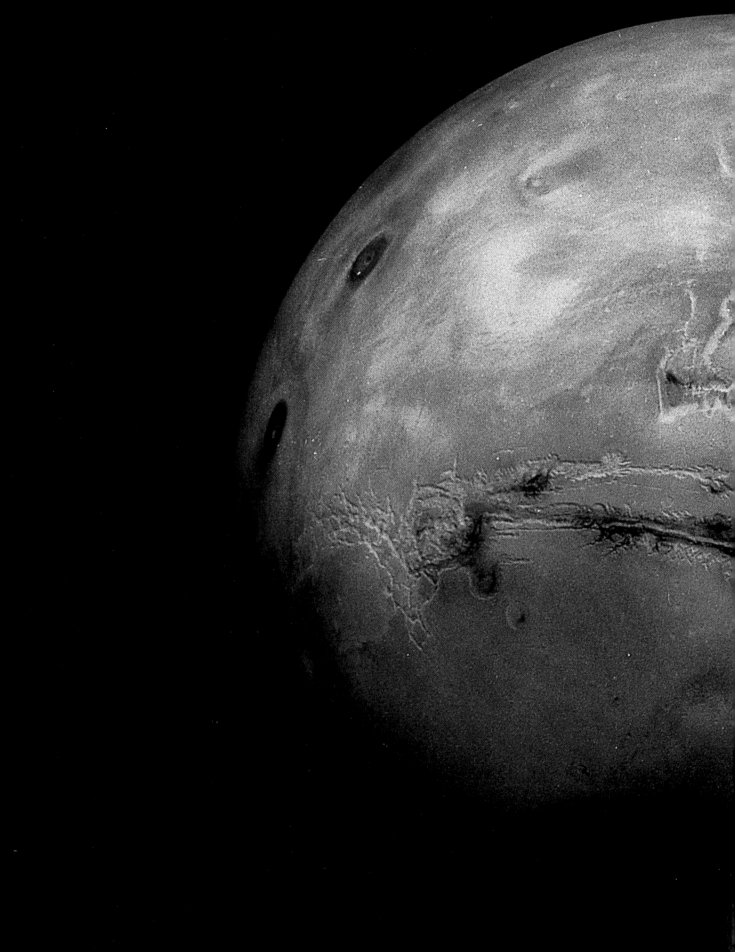

Jay Barbree & Martin Caidin
with Susan Wright

DESTINATION
MARS
IN ART, MYTH, AND SCIENCE

PENGUIN
STUDIO

PENGUIN STUDIO

Published by the Penguin Group
Penguin Putnam Inc., 375 Hudson Street,
New York, New York 10014, U.S.A.
Penguin Books Ltd, 27 Wrights Lane,
London W8 5TZ, England
Penguin Books Australia Ltd, Ringwood,
Victoria, Australia
Penguin Books Canada Ltd, 10 Alcorn Avenue,
Toronto, Ontario, Canada M4V 3B2
Penguin Books (N.Z.) Ltd, 182–190 Wairau Road,
Auckland 10, New Zealand

Penguin Books Ltd, Registered Offices:
Harmondsworth, Middlesex, England

First published in 1997 by Penguin Studio,
a member of Penguin Putnam Inc.

1 2 3 4 5 6 7 8 9 10

Copyright © Jay Barbree and
Martha A. Dee Dee Caidin, 1997
All rights reserved

The authors wish to acknowledge the following works as
valuable research sources in the preparation of this book:
Science Fiction: The Illustrated Encyclopedia by
John Clute, Dorling Kindersley Publishing, Inc.
Grolier Multimedia Encyclopedia of Science Fiction,
Grolier Electronic Publishing, Inc.

ISBN 0-670-86020-4
CIP data available.

Printed in the United States of America
Set in Cochin
Designed by Pei Loi Koay

In memory of

Martin Caidin

SEPTEMBER 14, 1927–MARCH 24, 1997

This is his last book,
the final chapter in his half century
of dedication to his devoted readers.

CAVU, Marty.

CONTENTS

DESTINATION MARS

IN ART, MYTH, AND SCIENCE

Chapter 1

THE MARTIANS ARE HERE!

N ASA has made a startling discovery that points to the possibility that a primitive form of microscopic life may have existed on Mars more than three billion years ago," a NASA press release stated on August 6, 1996, announcing the discovery of microfossils found in a meteorite from Antarctica.

The world's flagging attention suddenly returned to the subject of Mars, and a storm of speculations and statements rained down from experts, agencies, grassroots organizations, and even museums that housed Martian meteorites. Hardly twenty-four hours had passed before the National Space Society stated that political leaders and NASA should "initiate a program to send human explorers to the red planet."

Others, such as the Space Frontier Foundation, cautioned against a bureaucratic rush to Mars, urging the president to order NASA to "procure soil samples from private firms, which are better equipped to mount low-cost missions than the government."

1. The image of a crescent Mars reveals the southern view of the dawn side of the planet. The parallel lines of the great equatorial canyon system, Vallis Marineris, are faintly visible in the center, while near the top, hazy atmosphere clings to the protruding summit of the giant volcano Ascraeus Mons. (NASA/JPL)

John Pike, director of space policy for a think tank, the Federation of American Scientists, said that NASA must get moving on sending people to Mars, adding that whether there is life comparable to ours on Earth is "one of the most important questions confronting the scientific community."

Even without an agreement on exactly how to get to Mars, scientists have agreed that we should go there. Despite questions that were raised as to the validity of NASA's findings—are the tubular structures found in the meteorite truly microfossils of single-cell bacteria or not?—it is generally accepted that more samples of soil and rock are needed in order to confirm whether there is life on Mars.

Which fits nicely with NASA's plans. A little-regarded press release was sent out two weeks earlier, on July 20, with NASA announcing the projected launch of two spacecraft to Mars in November and December of 1996. Though the press release spoke of it being the twentieth anniversary of the successful Viking mission to Mars, people did not pay much attention until the flurry of interest started over the evidence "that life once existed on Mars."

So, once more, we are going to Mars. On the basis of the meteorite findings, NASA has announced that future missions to Mars will be revised to try to prove the existence of Martian life.

If some sort of primitive life is found, even a single-cell bacte-

2. This 4.5-billion-year-old rock is the meteorite in which fossil evidence was found of primitive life that may have existed on Mars. This meteorite was dislodged from Mars by a huge impact about 16 million years ago and fell to Earth, landing in Antarctica, 13,000 years ago.   (NASA)

ALH84001,0

 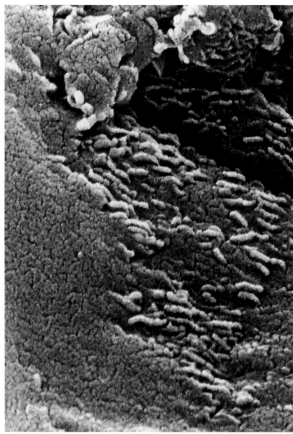

3. In the center of this electron microscopic image of a small chip from the meteorite are the tiny structures that are possible microscopic fossils of primitive, bacteria-like organisms. Scientists believe the organisms may have existed on Mars more than 3.6 billion years ago. (NASA)

4. This electron microscope image is a close-up of the center of the previous photo. A two-year investigation by a NASA research team found these organic molecules that are similar to microscopic fossils found inside Earth rocks. The largest structures are less than one-hundredth the diameter of a human hair, while most are ten times smaller. (NASA)

. . . .

ria, then that means our planet is not the only one to produce life. And if two planets in our own solar system have achieved that monumental step, then why not others in our galaxy? That takes us one step closer toward discovering advanced civilizations existing elsewhere on distant worlds.

The suddenly famous Mars meteorite with its tubular microfossils is not the first time Mars fever has gripped the general public, reviving a deep-seated cultural reaction to our neighboring planet. This feeling of uncertainty, of an almost overwhelming impulse to discover the truth, has prompt much of our observation of Mars since man first gazed into the sky.

5. Life on Mars: a comparison of the planets in our solar system (excluding Pluto). Mankind has always looked to Mars as the planet whose ecology could be closest to that of Earth's. The enormous gas giants would make Mars and Earth look like pinholes in a true-to-size comparative photo. (NASA)

6. As early as the seventeenth century, it was recognized that Mars had polar caps and rotated once every twenty-four hours, much like the Earth. Perception of the planet as Earth-like was subsequently reinforced by observations of white patches that were interpreted as clouds. In this photograph we see the relative size of Earth and Mars with a statistical comparison chart of the two planets. (NASA)

• •

COMPARISON OF EARTH TO MARS

EARTH		MARS
12 756 km	Diameter	6787 km
5.98×10^{24} kg	Mass	0.646×10^{24} kg
9.75 m/s^2	Gravitational acceleration	3.71 m/s^2
149.5×10^6 km (average)	Distance from Sun	227.8×10^6 km (average)
839 cal/cm^2/day	Sunlight intensity	371 cal/cm^2/sol
23° 27"	Inclination	23° 59'
24^{h}00^m	Length of day	24^{h}40^m (=1 sol)
365 days	Length of year	686 days (668 sols)
60 000γ	Magnetic field	50–100γ
1013 mb (average)	Atmospheric pressure	7 mb (average)
1	Known satellites	2

Behind this uncertainty lies fear—fear of the unknown, of things that could strike out and hurt us in ways we can't understand or predict. Fear of the "other."

Orson Welles capitalized on this fear, creating *War of the Worlds*, the most famous radio broadcast in history. On Halloween Eve, 1938, the Mercury Theater of the Air went live, with the actors performing under the guiding hand of Orson Welles.

War of the Worlds began innocently enough as a remote feed from a nonexistent hotel, covering the orchestra of Ramon Raquello. Almost immediately announcers interrupted the program to inform the American public that blasts of incandescent gas had been observed ripping away from Mars at specific and regular intervals. The Martians were coming!

Thus began Howard Koch's script based on the classic H. G. Wells novel *War of the Worlds* (1898). Koch crafted his invasion story in much the same way as H. G. Wells did in the original: shaking his head over the lost innocence of man, victim to

7. Orson Welles (at top, right) directs the Mercury Theater radio play of *War of the Worlds*. Backed by marvelous sound effects, his production swept into American homes and ignited a national panic. (CBS)

THE MARTIANS ARE HERE!

creatures much more powerful and terrible than we had imagined.

During the radio broadcast, Orson Welles and his team were mesmerizing, which was surprising both to Welles and to the audience. The actors and production staff transformed a marginal concept into a believable and terrifying drama.

It actually made sense for the audience to believe what they heard. In the 1930s, people relied on their radios for news and information, particularly information regarding the hostilities in Europe and the Pacific. Professionals in radio broadcasting helped instigate "war fever" with their reports. In the mid-thirties the Gran Chaco War was the scene of savage fighting in South America. The banana republics of Latin America hammered across the frontiers of neighboring lands. Japanese troops by the millions were pouring into and through China and were taking possession of French Indochina. A massive air war was being played out between Russia and Japan over the Manchurian wastelands. Civil war had already laid waste to great areas of Spain where fighting men of many different countries had joined in the struggle against fascism. And Hitler controlled half of Europe, with the Nazi tide continuing to spill across the continent.

Audiences were well aware of the possibility of sudden artillery fire and the roaring blasts of bombs falling on America. War was in the daily newspapers, on every radio newscast, in national magazines, and in theater newsreels. The hottest political struggles taking place in Washington focused on whether the United States would once again be dragged into the bloody morass of another European conflict.

8. The people who believed the *War of the Worlds* radio broadcast had no conception of the distance and amount of time it would take for Martian rockets to get here. Thirty years earlier an unnamed artist depicted a fleet of flying balloons reaching Mars (in the cover illustration of *La Guerre dans Mars*) only to enter into battle with Martians. It took minutes during the radio play to cross the 50 million miles, yet even at our current level of technology, it takes six months to fly to Mars.   (Mary Evans Picture Library)

So when the audience heard that powerful observatories around the world were training their sights on Martian ships approaching Earth, an invasion from Mars did not seem so far-fetched after all—particularly when there were on-the-scene reports from the great telescope centers (including some that didn't exist) confirming the original observations.

Orson Welles knew that his audience might misinterpret their radio play, and he insisted that CBS and its affiliated stations preannounce their performance. Warnings were repeatedly made during the show, identifying the writer as Howard Koch and the story as based on H. G. Wells's original *War of the Worlds,* published in serial form in *Pearson's Magazine* in 1897–98.

Yet many people didn't start listening until approximately ten minutes into the play, when the Mercury audience doubled from its usual 3.6 percent of the total to some 7 percent of all radio listeners. This happened when people turned away from the vocalist portion of the *Edgar Bergen and Charlie McCarthy Show,* and they suddenly found themselves listening to an emergency broadcast of a Martian invasion.

The CBS switchboard received hundreds of calls, demanding verification. Two years later, in 1940, Hadley Cantril wrote *The Invasion from Mars: A Study in the Psychology of Panic, with the Complete Script of the Famous Orson Welles Broadcast,* which reported on a series of interviews held by Princeton University a week after the broadcast. Cantril estimated that over a million listeners were frightened by the broadcast, with an additional ten million listening in breathlessly.

· ·

9. A monument marks the spot at Grover's Mill, New Jersey, where supposedly the Martians first landed. The inscription honors the 1938 broadcast of *War of the Worlds*: "This was to become a landmark in broadcast history, provoking continuing thought about media responsibility, social psychology, and civil defense. For a brief time as many as one million people throughout the country believed that Martians had invaded the earth beginning with Grover's Mill, New Jersey."
(Gerald Soffen)

To the audience, Mars suddenly became not so distant, in much the same way that the photographs of potential microfossils of extraterrestrial life have suddenly made Mars seem much more tantalizing.

WAR OF THE WORLDS

As if on celestial cue, fate added its own spectacular touch that Halloween Eve. Meteorites flashed through the night skies, and what usually would have been barely noticed now became evidence of Martian invaders.

Through a series of "interruptions," the radio play announced that powerful Earth shocks were being registered on seismographs in the vicinity of Princeton, New Jersey. Paul Stewart created the spectacular sound effects that contributed in a large part to the realism of the show.

In breathless tones, "reporters" described the alien cylinders that had smashed huge pits and gouges into the earth. One of

10. Martian machines land in the Thames Valley and march down the river on their spindly legs, pinchers prominently displayed and ready to snatch up fleeing Englishmen. This illustration by Warwick Goble is from the original version of H. G. Wells's story in *Pearson's Magazine* (1897).   (Mary Evans Picture Library)

11. During the radio play, one reporter fairly shrieked that he was watching things crawling and slithering from the cylinders, with the blazing eyes of monstrous creatures. The Martians had large bulbous bodies with writhing tentacles. From the babble in the background were cries of "Run for your lives!" amid the trampling rush to escape. (Mary Evans Picture Library)

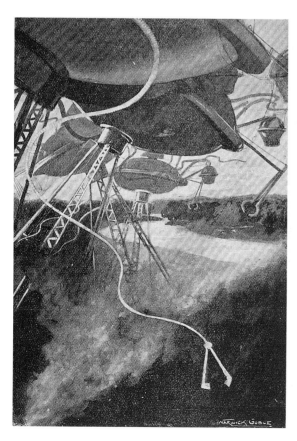

the actors, Frank Readick, based his performance on a recording made in 1937 of a reporter describing the descent and ultimate explosion of the *Hindenburg* airship.

Many of the audience heard the warnings that this was merely fiction, but they believed that the exhortations to remain calm were lies. They thought it was the government's way of controlling the situation.

The melodrama continued as one witness after another related the same heartbreaking scenarios of police, soldiers, and groups of farmers battling the Martians as they emerged from their ships. Yet rifles had no visible effect upon the attacking alien machines.

Then the military mounted a massive two-pronged attack with planes bombing the Martians with explosive missiles, while artillery fired at point-blank range. The Martians responded with flaming heat rays, setting everything from New Jersey to New York City on fire.

12. The first Martian emerges, confronting the first terrified human it encounters. Like H. G. Wells's original story, the 1938 radio play built the tension gradually, maintaining a realistic tone throughout. This illustration is from a French edition of *War of the Worlds* (1898).
(Mary Evans Picture Library)

13. The morning after his Halloween Eve rendition of *War of the Worlds*, twenty-three-year-old Orson Welles found himself surrounded by the news media. Welles told reporters that he had hesitations about this play because "it was our thought that people might be bored or annoyed at hearing a tale so improbable."   (CBS)

Only these Martians didn't just kill; according to the script, they "play with people like a cat plays with a dying bird or a mouse, batting them around like they were toys. The tentacles reach out from the machines, wriggling like huge snakes. They tear off arms and legs. . . ."

Martian war machines were landing all over the world, but just as there seemed to be no hope, the killing stopped. The corrosive effects of Earth's oxygen atmosphere disintegrated the Martians and their machines—exactly as H. G. Wells had described in his original story.

To those calm enough to notice, there were many telling dramatic devices, such as the passage of hours in a few minutes of radio time. The second half of the broadcast was much more noticeably fictional; after a brief program break, the story picked up a couple days after the initial invasion.

Day and night counterattacks were crunched into the second half of the show. Rust formed with unprecedented speed, taking mere minutes to reduce the mighty war machines to flaking hulks. Koch's version is notable for emphasizing that the Martians were not killed off by the most powerful military weapons in existence but by the lowest, most insignificant life forms on Earth—bacteria, viruses, midges, flies, and birds.

As Orson Welles read the last lines of the script, he added, out of character, a rather hasty admonishment to "remember the terrible lesson you learned tonight . . . and if your doorbell rings and there's no one there, that was no Martian . . . it's Halloween."

For the rest of the night, CBS put out hourly disclaimers, announcing that the invasion had been a fictional play. In Times Square in New York City, the Moving News sign stated: "Orson Welles Frightens the Nation." The New York *Daily News* rushed its newspapers to the streets to allay panic. "The Halloween Broadcast," wrote George Dixon, "created almost unbelievable scenes of terror in New York, New Jersey, the South, and as far west as San Francisco."

Yet Orson Welles was more surprised than anyone that people had believed the show.

14. Many newspapers rushed "Extras" to the streets to allay panic and to convince readers that the only real danger was their own hysteria. This newspaper account appeared forty-four years after Orson Welles's radio broadcast.
(Asbury Park Press)

The night Martians invaded New Jersey

Orson Welles' ruse

He had not liked it during rehearsals and had kept pushing for more realism. He had been afraid that the audience would find the play boring, and when he realized the fervor that was being caused, even before the show ended, he tried to palm the whole thing off as a Halloween prank.

Welles was both praised for his "contributions to the social sciences" and condemned for not having seen "the effect of too much dramatic realism on an audience already strung to high nervous tension." Legal actions were brought against CBS and Mercury Theater, but none succeeded. What the *War of the Worlds* broadcast did do was make Orson Welles famous. And it revived the ancient fear of Mars in our culture.

15. This anniversary story about the "invasion" appeared in the 1972 *Asbury Park Press*, describing the thousands of people who rushed to where the Martians had landed. An insert picture is captioned: "Legend has it this Grover's Mill windmill drew shotgun fire as a Martian warship."
(Artist rendition by James Denk, *Asbury Park Press*)

THE MYTHOLOGICAL ARCHETYPE

What made Welles's broadcast of a Martian attack seem so eerily real? Why does the idea of invading Venusians or Plutonians seem incongruous, while the name "Martian" rolls off our tongues with ease? Perhaps it is because we have always thought of Mars in terms of war and technological power.

Mars today is not so readily associated with the ancient Greek myths of battles backed by all-powerful gods. Yet mythological archetypes are the foundation of our modern beliefs. Just look at our heroes in the form of celebrities, sports players, musicians, and movie stars. There you find our modern-day Hercules and Odysseus and Helen of Troy.

It is part of our cultural heritage that we associate Mars with war. The Greek god Ares was the patron of passion and battle, and the Roman cult of Mars immortalized the red planet, offering sacrifices to the god who made it possible for the Romans to dominate the Western world.

Myths explain the unexplainable. They are the stories of the miraculous, the intuitive, and as such are not expected to represent science, something that can be proved or verified. Myth fills in the gaps between knowledge and fear, giving people a way to understand the uncontrollable aspects of life.

Even though people today rely on laws of science to explain reality, we continue to put equal weight on subjective explanations. For proof of that, simply look at the First Amendment of the American Constitution. That first law uphold the freedom of speech, supporting the right of individuals to hold differing perspectives on reality.

In order to find out how the myth of Mars has permeated society, we must look to the ancient world, when early astronomers first looked to the sky.

16. *War of the Worlds* was remade as a movie in 1953 by producer George Pal. H. G. Wells's classic of alien invasion has been perhaps more copied than any other sci-fi tale in history. This scene in the poster shows how Well's original version was updated, with flying saucers attacking Earth instead of the ungainly walking tripods. Yet the invasion myth has endured with such strength that the plot needed little changing to make Wells's story acceptable to modern audiences. (Archive Photos)

Chapter 2

THE MYTH OF MARS

M ars has teased the imagination ever since the first astronomers, unnamed and lost in the dust of history, began serious study of the night sky. Unlike the countless other lights in the sky, Mars doesn't flicker. And instead of a white light, it glows red.

Humanity's ancient understanding of the movements of the stars can be seen in the construction of enormous monuments. Stonehenge was a celestial timekeeper, built in England around 2000–1500 B.C. Earlier, around 2500 B.C., the Egyptians erected three pyramids at Giza. The Cheops Pyramid was aligned to the polestar, and the seasons can be read in the position of the pyramid's shadow. Experts have claimed that the positions of the three pyramids reflect the slightly offset position of the three stars in Orion's belt.

The Sumerians also studied the stars during the middle of the third millennium B.C., as both an aid to navigation and for time-

17. This photograph of Mars was taken thousands of years after the first astronomers began observing the red star in the sky. Long before the planet became identified as Earth's neighbor and a potential home for other civilizations, the ancient myths considered Mars to be the god of war. (NASA)

• •

keeping. They were the first to create almanacs, rather than monumental markers, to record the best times to plant and harvest crops.

These ancient astronomers realized that the positions of the stars remained relative to one another and were consistent in their movements across the sky. Other celestial bodies came and went; meteors blazed through the atmosphere, and comets dragged their gossamer tails across the sky for days or even weeks. There were also lights that ghosted mysteriously though the night, awing millions who believed they were witnessing the passage of a human soul on its way to some heavenly reward.

But Mars was unlike the other "fixed" stars. It seems to perform a loop-the-loop in the sky, moving forward and backward in odd, yet predictable, patterns. Five thousand years ago, the Egyptians recognized the retrograde motion of Mars, calling the planet *sekded-ef em khetkhet,* meaning one "who travels backward."

Ancient observers also noted a periodic change in the brightness of Mars. It would be thousands of years before astronomers learned that this variation was caused by the elliptical orbits of the planets around the Sun.

BABYLONIAN ASTRONOMY

The ancient Babylonians were undoubtedly the most capable and precise astronomers among the ancient cultures. After centuries of watching and recording, they were amazingly adept at cataloging not only individual stars but also their positions in the sky for any given time of the year. And among the most reliable signposts was Mars, bright and red, especially when the planet was low in the sky.

Babylonian astronomy dates back to about 1800 B.C., coinciding with the formation of Babylonian culture on the banks of the Euphrates during the Bronze Age. Their observations of the movements in the heavens led to the first systematic, scientific treatment of the physical world. The Babylonians used the Egyptian calendar to chart the periods between observations and created the zodiac as a reference system for the sky.

Though they were able to predict the positions of both the stars and planets in the sky (even the Earth and the Moon), the Babylonians never developed mathematical or geometrical theories on which these movements were based. During this period of stargazing and wondering, celestial objects were studied as if they were lights suspended magically on the solid vault of the heavens.

In this way, cosmic activity such as meteor showers, which returned with annual regularity, were soon acknowledged with proper religious ceremonies and festivals. Priests were the possessors of the knowledge of the stars, and these spiritual astronomers guided the behavior of Babylonian society according to what they saw in the juxtapositions of celestial bodies.

GREEK ASTRONOMY

The Greek word *planetes* means "wanderers," acknowledging the erratic movements of the planets. Their circuit of the zodiacal constellations takes anywhere from eighty-eight days for Mercury to twenty-nine and a half years for Saturn. The inner planets move more rapidly in their orbits, thus Mercury was named for Hermes, the fleet messenger of the gods, because it goes the fastest. Venus, Earth, and Mars move progressively slower, while the outer planets pass through our night sky with stately leisure, as befitting the dominant gods of Greek mythology, Zeus and Neptune.

Though the Greek myths could supply reasons for the erratic movements of the planets, those reasons did not satisfy observers for very long. The Greeks gave humanity a scientific, skeptical philosophy—one of experimentation based on mathematical principles. Their main science was geometry, the "measurement of the Earth," using methods employed by Egyptian rope-stretchers in their annual surveys, reestablishing the boundaries of farmlands flooded by the Nile. The Great Pyramids were built using the expertise of these rope-stretchers with measurements based on the "cubit," the length of a man's arm, but the Greeks used geometry to study the universe.

Greek observers recorded the complicated motions of the wandering planets, tracking the variations that were partly caused (unknown to them) by the Earth's wobble on its axis. The complicated motions of these planets helped generate further interest in the heavens. If the motions had been simple, they could have been explained in reductive terms by mythology.

The Pythagoreans (fifth century B.C.) created the first of the Greek astronomical theories—that man could understand the harmonies of the universe by contemplating the regular motions of the heavens. They believed there was a central fire around which all celestial bodies revolved.

THE GREEK MYTH OF ARES

As with all ancient astronomy, the movements and changes could be recorded but not explained. Thus, they became the objects of myth. In every culture on Earth, the sky is considered sacred and related to the highest divinity.

The Greeks were the first to partially separate cosmology—the study of the sky—from mythology. Yet both remained inextricably entwined for most of human history. Science only slowly escaped the chains of mythic belief; even Kepler and Newton based their theories on metaphysical beliefs involving eternity, infinity, and the omnipotence of God.

Under the Greek concept of time, the end of the world would come whenever Mars, Earth, and the other planets came into conjunction—that is, when they lined up together. The catastrophe that would be unleashed would destroy the old cycle and initiate a new one. The length of this cycle was unpredictable, similar in its infernal nature to that of Mars itself. This mythical, cyclical view can still be found in the modern theory of the big bang: an expanding universe that eventually comes to a halt, collapsing in on itself, only to result in another big bang.

The Greeks considered their gods, particularly the twelve Olympians, to be immortal controllers of natural forces, and

18. The Romans copied the Greek culture in both their art and their religion, as can be seen in this Roman copy of a Greek statue by Alkamenos during the fifth century B.C. Like many other legacies from the city-states of Greece—such as politics, philosophy, and science—the myth of Mars has endured to the present day.
(Erich Lessing/Art Resource, NY)

thereby explained things that could not be understood. Each of the gods took on a particular attribute: Ares (called Mars during the Roman period) was the god of war.

With Ares, it is easy to see how the foundations of myth were firmly based on the physical, systematic observations of the sky. The unusual motion of Mars is unlike that of any other planet, and this led to suspicion and distrust. Ares became the embodiment of unreasoned disorder—fighting, killing, and passionate lust—on an astronomical scale.

Like so many other legacies bequeathed to Western culture by the Greek civilization, our concept of Mars is forever linked to aggression and strife. We have difficulty recognizing our own cultural myths because they are taken for granted, serving as the basis for our questioning of the universe (this curiosity is also cultural, passed on by the Greeks). The success of Orson Welles's radio play of *War of the Worlds* revealed how powerful myth can be; the realistic invasion would not have been so readily believed if it had been an attack by Venusians instead of Martians.

Myths have power because they are seen as fundamental truths outside the realm of human experience. Myths are not called upon to justify the facts or prove themselves. In fact, the Greek word *mythos* means "word," as in a final pronouncement on both natural phenomenon and human morals. This differs from the Greek *logos*, in which validity or truth can be argued and demonstrated.

People try to understand myths in order to know how to behave in society. Since the Greeks valued reason and reflection above all, Ares was seen as the embodiment of negative virtues. Only cultures based on martial law could appreciate Ares, such as the Greek city-state of Sparta and, later, the Roman Empire.

19. The fresco of Mars and Venus, from the Casa dell'Amore *punito*, Pompeii, shows the dual sides of the passionate Ares—god of battle and seduction. Though Ares was not a popular god among the Greeks, his myth served an important function, symbolizing the uncertainties of human existence, such as life and death, war and peace, love and hate. Because of his attributes, Ares was both feared and despised.

(Erich Lessing/Art Resource, NY)

20. According to Homer's *Iliad*, Ares is accompanied into battle by his sister Eris (Strife). She drives the chariot in this fresco, while Diomedes casts his spear at Mars. Diomedes was one of the most respected leaders in the Trojan War, and at one point he even wounds Aphrodite.

(Archive Photos)

HOMER

One of the main sources of our knowledge of Greek myths comes from Homer's epic poems *Iliad* and *Odyssey*. Ares figured prominently in both epics in the role of villain. It has been said that Homer wrote the *Iliad* in order to show why Ares, the god of war, was justly the most hated god among his family on Olympus. In the Roman era, Venus (the goddess associated with Aphrodite) was the wife of Mars, or Ares. Their children—Deimos and Phobos, better known as Panic and Rout—were the gods who accompany Ares into battle, along with his sister Eris.

21. Traditionally the relationship between Mars and Venus is considered to be an allegory of Strife overcome by Love. This painting by Titian during the Italian Renaissance in the fifteenth century reveals the permanent link the ruling class made between the affair of Mars with Venus and their code of chivalry. The two gods were often depicted reclining together while cupids played with Mars's idle weapons of war.

(Erich Lessing/Art Resource, NY)

· ·

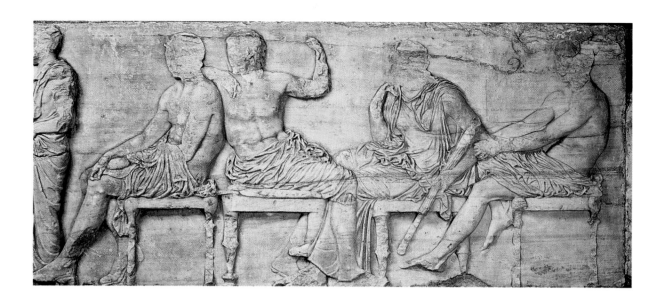

22. This group from the east frieze of the Parthenon (477–432 B.C.) shows Ares, god of war, seated on the right. He leans back in a casual pose, wearing loose drapery rather than his typical short battle skirt. The classical qualities of idealization dictated that the figure of Ares be harmonized with the other gods: from left to right, Hermes, Dionysus, and Demeter next to Ares.

(The Bridgeman Art Library/Art Resource)

23. This bronze statue of Mars reveals the profound interaction between the Etruscans and the Greeks during the Archaic period, sixth century B.C. The Etruscans adopted much of the Greek pantheon, with some of their original gods made equivalent to Greek gods, such as Tin for Zeus, while Apollo and Ares were directly annexed. In this Etruscan statue, Mars wears the traditional Greek battle dress.   (Art Resource, NY)

THE CULT OF ARES

The Greeks, who prospered under democratic rule in the distinct city-states, carried their religion and culture with them as they embarked on trade and colonization. The cult of Ares began to spread during the eleventh century B.C., and his worship lasted along with most of Greek religion under Roman rule until the reign of Emperor Julian (fourth century A.D.).

In Greece, Ares was primarily worshipped in Sparta, the ancient capital of Laconia, on the southeastern Peloponnesus. There is little documentation of the cult of Ares in Sparta. The scarcity of ruins and art reflects the austere Spartan culture, with the military hierarchy directly opposing opulent public works.

In Athens there was a temple at the foot of the Areopagus, or hill of Ares. This hill is the legendary location of the murder trial of Ares and the traditional location of homicide trials. At Geronthrae in Laconia, no women were allowed in the sacred grove during the festival of Ares, while at Tegea he was honored by a woman's sacrifice and was called Gynaikothoinas (entertainer of women), possibly alluding to his passionate and procreative side, which had proven so irresistible to Aphrodite, the goddess of love.

24. This relief from the north frieze of the Siphnian Treasury (525 B.C.) depicts Ares during the battle of the giants against the gods. The temple was built in Delphi to honor Apollo, erected by the people of the island of Siphnus. The frieze decorated all four sides of the temple and is one of the best surviving examples of the Archaic style of sculpture. The shallow reliefs are little more than drawing in the round, with depth created by overlapping figures. (Erich Lessing/Art Resource, NY)

By 400 B.C., the Greeks understood that Earth was spherical, based primarily on the shape of the shadow it casts on the Moon during a lunar eclipse.

Plato was responsible for the ensuing Greek struggle to make astronomical observation agree with the ideal spherical universe. Platonic ideals of harmony dictated that geometry was the abstract guideline of the true, perfect universe, while the real world was but a pale, imperfect reflection. For almost two thou-

25. This engraving shows Hipparchus (ca. 130 B.C.) pinpointing the gleaming red eye of Mars and the dazzling light of Venus from the observatory in Alexandria. Nearly all of Hipparchus' writings were lost, yet his scientific observations were recorded by Ptolemy. Hipparchus is believed to have created the first catalog of the stars, showing their brightness (magnitude) and position.

(Archive Photos)

sand years after Plato, astronomers continued to try to create models of the heavens erroneously based on concentric spheres.

Eudoxus (b. ca. 400 B.C.) was the first to create a model of the universe according to the Platonic ideal. He believed the model should be accurate and internally consistent, as harmonious as a song or poem. Eudoxus theorized there were twenty-seven spheres around the Earth that affected one another, altering the paths and velocities of the planets and stars.

ALEXANDER THE GREAT

The mathematics of Eudoxus' model could not hold up to the numerous Babylonian astronomical records that were brought to Greece by Alexander the Great in 330 B.C.

Alexander the Great was a pupil of Aristotle, and his own interest in scientific inquiry helped nurture advances in geography and natural history. He conquered the Near East and split the Greek world into territorial monarchies. The knowledge of the classical Greek philosophers and artisans was spread throughout a cultural empire stretching from Gibraltar to India. On this domain, the Roman Empire was later built.

As skeptical experimenters disproved Eudoxus' calculations, it was left to Alexander's teacher, Aristotle, to develop a new theory based on the old model. Yet this theory, put forth in Aristotle's book *De Caelo (On the Heavens)*, was based on one fundamental flaw: he placed the Earth at the center of the universe. Most subsequent theories were based on Aristotle's model and were forced to become ever more complicated as they tried to explain the observed motions of the Sun, the Moon, planets, and stars across the sky.

CLAUDIUS PTOLEMY

Alexander established numerous cities in his own name. Alexandria in Egypt was created specifically to be a capital of learning modeled on the Greek ideal. And indeed it became the center of intellectual life in Egypt under the Ptolemies.

In this city, the Greek ideal of an Earth-centered universe of nested spheres was immortalized by Claudius Ptolemy (working around A.D. 140). Ptolemy's system stripped the Aristotelian model of its symmetry in an attempt to explain the complicated motion of the heavenly bodies. He incorporated epicycles and eccentrics—as if planets and stars were rotating in circles around the line of their movements. And some of the spheres were not perfectly centered but slipped to one side. This was all mathematical sleight of hand in order to make the elliptical orbits of the planets seem to behave like circles.

Ptolemy admitted that his model was awkward when it came to the sequence of motions, protesting, "But when we study what happens in the sky, we are not at all disturbed by such a mixture of motions."

Ptolemy's thorough compilation of earlier Greek mathematicians and astronomers made his book, *Almagest*, the basis of astronomy for the next millennium. It would not be until the scholars of the Middle Ages determined the true enormity of our solar system that Ptolemy's model would begin to break down.

ROMAN CULT OF MARS

For hundreds of years the god Mars dominated the Roman Empire, the leading civilization in the Western world. Founded in central Italy, the Roman Republic grew from a small city-state to an empire that dominated Europe, the Mediterranean, and the Middle East.

The Roman Empire, always run by authority figures, gave rise to a cultural tendency to obey rather than question orders. This was quite the opposite of the Greek love for impartial experimentation, and it led to a stagnation in the sciences. The

26. This frieze is from the Altar of Domitius Ahenobarbus (ca. first century B.C.). Ahenobarbus was one of the chief Roman generals for Mark Antony after the assassins of Julius Caesar were defeated. Ahenobarbus opposed the Egyptian queen Cleopatra's dominance over Antony and deserted to Octavian shortly before Antony was defeated. The Ahenobarbus family was notorious for their pride and cruelty, and they built this temple to Mars in order to gain the god's support.

(Giraudon/Art Resource, NY)

Romans were a practical people, gathering useful knowledge from the cultures they conquered in order to live better in their material world.

Roman citizens considered Mars to be their protector, and his followers eventually outnumbered those of Jupiter. Mars was said to be the father of Romulus and Remus, the legendary founders of Rome. He was thus attributed with a certain paternity at the birth of the Roman state. His female counterpart was his sister Bellona, the Roman goddess of war, usually depicted as a matron in armor.

The spread of the Mars cult was strong even in distant lands, because he was considered to be the spiritual commander in chief of the Roman army. As the legions spread, so did the myth of Mars. After battling the Romans for decades, the chieftains of the fierce and combative Germanic tribes decided they could benefit too by paying homage to the Roman gods. They began to

27. Roman frescoes were painted inside private homes and were intended to create an illusion that there were no walls. As in this fresco (first century B.C.) of the wedding of Venus and Mars, painters often incorporated architectural features such as columns and windows into their scenes. Many of the figures were shown life size.

(Alinari/Art Resource, NY)

28. This fresco of Mars and Venus is from Pompeii, one of the cities destroyed by the eruption of Vesuvius in A.D. 79. The remains were buried until the eighteenth century, perfectly preserving a record of Roman life and culture. The Romans added fresh vigor and realism to their images of Mars.　(Erich Lessing/Art Resource, NY)

imitate Roman rituals, and Ziu, the Mars god of the Germans, began to hold sway over the people. Ziu's powers had always been respected, but enhanced by Roman influence, the god became a renewed symbol of strength.

Until Augustus (27 B.C.–A.D. 14), Mars had only two temples in Rome: one in the exercising ground of the army, the Campus Martius (ca. 160 B.C.), and the other outside the Porta Capens. But during his reign, the worship of Mars gained new impetus. As Mars Ultor (Mars the Avenger) he became the personal guardian of Augustus in his role as avenger of Caesar. At a shrine within the city sacred spears were kept representing Mars Hasta (Mars the Spear). Whenever war broke out, the consul shook the spear saying, "Mars vigila!" (Arise Mars!) The temple of Mars Ultor in the Forum of Augustus in Rome was built at the height of Hellenistic influence (ca. 25 B.C.). The ruins of the temple can still be seen in Rome.

LIFE ON OTHER PLANETS

A transition began to occur during the second century A.D., reflecting the fact that the citizens of the Roman Empire were facing an uncertain future. In the art of the day, the serenity and harmony of the figures were replaced by angular features and agitated gestures, reflecting the horror and tragedy of constant warfare.

Out of this tumultuous time came the first speculations on extraterrestrial life. Adhering to the Roman tradition of practicality and realism, Plutarch (ca. A.D. 46–120) recounted the tale of a man using wings to fly to the Moon, where he found a race similar to our own.

Lucian (ca. A.D. 120) picked up the space travel idea in his *Vera Historia*, but he added something new to the budding genre. As history shows, in times of major cultural transition, authors often respond with satirical stories designed to reflect the follies of their own decaying society.

In *Vera Historia*, Lucian describes an absurd lunar war involving sixty million infantry, three-headed birds, venomous spiders hundreds of feet long, not to mention garlic throwers, three hundred thousand flea riders, and swarms of ants bigger than ele-

29. In *Vera Historia*, a fictional story of space travel written in the second century A.D., Lucian describes a sailing vessel that is lifted from the sea by a violent cyclone and carried to the Moon. There, the hapless adventurers explore and witness a war before turning their ship and sailing back to Earth. (Drawing by James Akins)

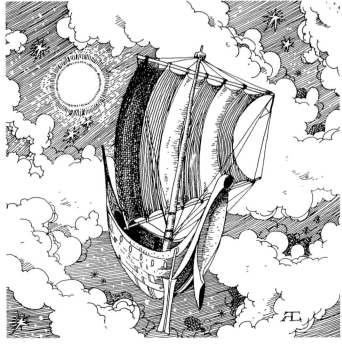

THE MYTH OF MARS

47

phants. Very likely, his readers never looked on the Moon in quite the same way again.

Like Orson Welles, who pushed his writers, actors, and technicians to create a realistic program, Lucian presented his material in a such a way that his audience was left dumbfounded, considering the real possibility of travel to other worlds. Like Welles, Lucian gave warning, saying that he was a liar and hoaxer. This open avowal similarly led to suspicions that the tale was somehow true. Why else would he try so hard to convince everyone that it wasn't? So goes popular logic.

Chapter 3

CHRISTIANITY AND ARIES
IN THE MIDDLE AGES

The successor to the great Roman Empire was Christianity. In A.D. 313 the Roman emperor Constantine instigated the decline of cult worship and the ascendancy of the Church by converting to Christianity. The Church's attitude to scientific inquiry was, at best, hostile. The earth was seen as flat and the sky belonged to Heaven and God, and nature could not be studied or questioned. As Saint Ambrose said in the fourth century, "To discuss the nature and position of the Earth does not help us in our hope of the life to come."

It is even alleged that Christian zealots burned the pagan books in the library of Alexandria, wiping out a venerable institution of learning and philosophy and much of ancient scholarship. Scholars fled from Alexandria and Rome and sought refuge in Byzantium. But Constantine conquered Byzantium and converted the people to Christianity, renaming the city after himself — Constantinople.

Christianity didn't have much use for Mars until relatively recently. In 1938, C. S. Lewis wrote *Out of the Silent Planet*, a sophisticated Christian rebuttal against the use of science for material gain. Lewis set the action on Mars, attacking H. G. Wells in particular, whom he portrayed as a vulgar journalist.

Typical of Christian apologists, Lewis set himself in opposition to scientists and government research. He described the guiding spirit of Earth as a twisted, fallen creature very much in the spirit of fantasy. Yet he conscientiously considered the scientific effect of weaker gravity on Martian plants and animals.

Forty years earlier, James Cowen, one of the first apologists, wrote *Daybreak: A Romance of an Old World* (1896), featuring a trip to Mars. There, parallel evolution had created another Christianity with another incarnation of Christ.

CHRISTIANITY AND SCIENCE

For more than twelve hundred years following the tales of Lucian and Plutarch, no writer dared be guilty of heresy by claiming that more than one world with living beings existed. This would have been seen as a direct contradiction of Church doctrine, as based on the Scriptures.

Naturally, science fared badly under this dogmatism, and the Ptolemaic model stood unchallenged for centuries despite cosmologists' increasing awareness that Ptolemy's predictions did not mesh with observed phenomena.

The greatest scientific gift Christianity gave to modern man was a change in the conception of time. The belief in cyclical time had kept the Greeks from inquiring too seriously into the true extent of the past. The cycle was seen as eternal and, for the most part, unchanging except in its details.

Christianity replaced this cyclical construct with a linear form of time, which had a definite beginning, middle, and end, as testified by the Scriptures. According to the Old Testament, the list of "begets" detailed the march of time through the ages, from Adam to Moses, down to the House of David.

Ironically, this narrow approach to time instilled the idea that the universe had a beginning and that Earth and everything on

30. The image of Mars began to recede during the Middle Ages, when the stars were seen simply as God's lights in the sky. The red planet, however, took on another aspect in astrology and was considered to govern people's courage and initiative in much the same way Mars, god of war, represented glorious vigor and sacrifice. (NASA)

it could be subject to dating. This opened the door for the scientific inquiries that would follow during the Renaissance.

Under this concept of time, human affairs assumed a prominent position according to how they pertained to God—exalting man over the physical Earth and universe. This focus on individual actions helped nurture the rise of astrology, which directly related man's actions to the movements of the heavens.

31. This view of the Sistine Chapel before Michelangelo painted his famous frescoes shows the ceiling painted dark blue and patterned with stars. From the early Christian era, the Church viewed the universe as a tent with the stars and planets pushed around by angels—an explanation that easily accounted for the odd stellar and planetary behavior.   (Alinari/Art Resource, NY)

Celestial omens were first categorized in ancient Mesopotamia, foretelling success or disaster. This led to a serious study of the positions of the stars and the possible meaning of their interrelationships. The zodiacal system was developed as a numerical reference scheme around 450 B.C. when the ecliptic belt was discovered.

The ecliptic belt is referred to in both astronomy and astrology as the zodiac. The zodiac contains all the motions of the observed planets and extends to nine degrees on either side of the Sun's apparent annual path. This belt was divided into twelve constellations and signs, each occupying one-twelfth, or thirty degrees, of the circle.

The oldest record of the zodiacal signs is from Babylon, a cuneiform horoscope from 419 B.C. In the third century B.C., the zodiac reached Egypt where it was merged with the system of decans—the thirty-six bright stars or star configurations spaced at approximately equal intervals from east to west around the heavens. Each was considered a dominant god or influence for about ten days of the year.

The most famous Egyptian star map is from the first century B.C., a stone chart found in the temple at Dandarah and now at the Louvre. The Zodiac of Dandarah illustrates the Egyptian decans and constellations and incorporates the Babylonian zodiac as well, with many of the stars doubly represented. Naturally, this zodiac is hardly an accurate rendering of the heavens.

It was the Ptolemies (a Greek dynasty ruling from 305 to 30 B.C.) who put mathematics to the task of calculating the connection between the constellations and man. The theoretical basis of Greek philosophy dictated that physical fact was based on an ideal interpretation. Special relations were believed to exist between celestial bodies and their varied motions. The complexities of these relations were believed to reflect the complexity of the world.

Much like myth, astrology is a method of explaining events and is based on the assumption that the planets and stars,

32. The study of astrology professes to interpret the positions of the planets and how they affect human affairs. This ceiling fresco (Italian, fifteenth century) of the constellation Aries reveals its associations with the Mars god of war and fertility. Aries came to symbolize the human capacity for change and growth, as well as courage to face death.
(Alinari/Art Resource, NY)

when considered in their movements and configurations (called constellations), can indicate changes in personal activities. These relationships can be calculated from the date of birth, covering the course of a person's life until death.

Thus, the myth of Mars continues from antiquity in the form of astrology. Mars is the first house of the zodiac, and is associated with Aries (formerly it was Scorpio, which is now considered to be ruled by Pluto).

ATTRIBUTES OF MARS

A variety of attributes of the human character are associated with the signs. The house of Mars, zodiacal sign Aries, is considered to represent masculine, aggressive attributes, primarily concerned with self-preservation.

It is also said that people born under the sign of Aries have a strong urge to set themselves apart from others, to emphasize their own individuality, and to prove their own worth. They express themselves ardently, forthrightly, and they take brave action based on strong desires, such as sexuality and adventurousness. Sometimes this impetuosity may lead to lack of self-restraint, even destructiveness and intolerance. When the so-called Martian drive is weak, the individual is timid and lacks energy and ambition.

The Greeks called the zodiac the *zodiakos kyklos*, or circle of animals. Aries, characterized by the ram, is identified with both the Egyptian god Amon and the Greek myth of the ram with the golden fleece.

33. The sign of Aries, six scenes. This fresco is in the Palazzo della Rugione in Padua, Italy. Astrologically, Mars symbolizes the energy we direct out to the external world—the raw, brute force ruled by passion rather than intellect. Aries lasts from March 21 to April 19.

(Scala/Art Resource, NY)

Manilius (ca. 30 B.C.–A.D. 37), an astrologer who lived during the reigns of Augustus and Tiberius, described a Mars-type person positively in his *Astronomica.* "Should his aggressiveness be needed in a righteous cause," Manilius wrote, "depravity will turn into a virtue and he will succeed in bringing wars to a conclusion and enriching his country with glorious triumphs."

Mars is considered to be at its worst when it is found in hard aspect to a personal planet such as Saturn, Uranus, Neptune, or Pluto. Sibley, an eighteenth-century astrologer, insisted that, "If Mars is ill-dignified [found in Taurus, Libra, or Cancer] . . . the individual will then be a trumpeter of his own fame, without decency or honesty; a lover of malicious quarrels and affrays; prone to wickedness and slaughter; . . . of a turbulent spirit, obscene, rash, inhuman, and treacherous, fearing neither God nor man, given up to every species of fraud, violence, cruelty, and oppression."

Adolf Hitler had Mars in Taurus, and astrologers see his tactical mistake in hesitating to invade England as an example of the stubborn, intractable trait of this detriment. Hitler was also obsessively stubborn about pursuing his Russian campaign, flying in the face of all reason as well as the historical precedence of Napoleon's failure when confronted by the Russian winter.

John F. Kennedy also had Mars in Taurus, and he displayed a similar hesitation during the 1961 Bay of Pigs invasion, when he failed to support Cuban rebels who were prepared and outfitted by Americans. Fidel Castro subdued the rebels without the promised U.S. air support, which Kennedy acknowledged was a mistake. Yet a year later, during the Cuban missile crisis between Russia and the United States, Kennedy displayed his Martian stubbornness, refusing to allow Premier Khrushchev to build his missile bases from which it would have been possible to attack U.S. cities.

THE CHURCH AND ASTROLOGY

A stringent philosophy, astrology defines a determined universe, denying free will; this lies counter to the philosophy of the

Christian world. Yet in the interpretation of Bardesanes, a Syrian Christian scholar (A.D. 154–222), the motions of the stars govern only the elemental world, leaving the soul free to choose between good and evil. The Christian Priscillianists, followers of a Spanish ascetic of the fourth century, Priscillian, saw the stars as a way of understanding the will of God—but only to those training in astrological symbolism. Basically the Church was not averse to symbols and myths. The Old Testament was based on mythological material, and the Platonic dualism of physicality versus idealism was transformed into Christianity's view of an imperfect body and an immortal soul.

34. This manuscript illumination of Aries the Ram is from an Italian Book of Hours (fifteenth century). Manuscripts were made of parchment, thin pieces of animal skins, and the pigments were mixed from rare minerals, including gold. Monks in their scriptoriums would work for years on one manuscript, etching in tiny details on miniature scenes that were often under five inches square.

(The Pierpont Morgan Library/Art Resource, NY)

The rituals of polytheism can also be seen in the festivals and processions of Christianity. The numerous holy saints took the place of lesser, local deities, while the shrine to Mary at Lourdes can be seen as a direct counterpart to the pagan healing shrine at Epidaurus in Greece.

GREEK TRANSLATIONS

It was left to the Islamic scholars to study and elaborate upon the classics of Greek science. During the last surge of paganism in the fifth and sixth centuries A.D., Byzantium (the Eastern Roman Empire) gave birth to an abundance of astrologers: Hephaestion, Julian of Laodicea, Proclus, Rhetorius, and John Lydus. But the decline of intellectual life in Byzantium after the sixth century persisted until the eighth century, when translations of Hellenic and Syriac texts once again began to appear.

Al-Farghani, a ninth-century astronomer, established the scale of the solar system based on Ptolemy's model. Al-Farghani figured that the spheres fit tightly together — "there is no void between the heavens" — claiming that it was one-tenth its believed size. Thus, Islamic astronomers helped undermine their beloved Ptolemy by rendering the abstract spheres into concrete numbers.

Translations into Arabic spread to Spain and Sicily in the twelth and thirteenth centuries, and the greatest flowering of astrology came during the late fourteenth century when John Abramius and his students revised the older astrological treatises in Greek for Renaissance scholars. For the next two centuries, astrology flourished.

This revival in Greek learning also led to a wealth of new translations on science and astronomy. Most of the Greek scientific texts had been preserved in Arabic translations by the Islamic scholars from the seventh century onward. The presence of a large Muslim community in Spain contributed to the spread of these texts during the Middle Ages, helping to lay the groundwork for a scientific revolution during the Renaissance. By then, the mechanical developments of the Middle

Ages (particularly in agriculture, mining, and the printing press) aided in the advancement of science.

EXPLORATION

Navigation served to revive an interest in real celestial phenomenon, as the great European explorers from Marco Polo in the thirteenth century to Columbus in the late fifteenth century set off to expand their nation's reach and make a fortune for themselves.

In order to return to where they started, explorers had to be able to know where they were. Columbus didn't persuade Queen Isabella of Spain to finance his expedition by promising to prove the world was round; he convinced her it was much smaller than scholars had estimated. He believed he could get from the coast of Spain to the East Indies faster by sailing west.

Isabella's court geometers insisted that it would take three years to sail Columbus's route, and they would have been right if not for the existence of the western continents. Nearly thirty years after Columbus reached North America, Ferdinand Magellan circumnavigated the Earth, taking three years for his journey.

But it was the bravery of these great explorers who encouraged westerners to think of the Earth in more accessible terms. This was true of the heavens as well.

In 1277, Pope John XXI granted the Bishop of Paris permission to state that there could be more than one world in space. The Church believed that to say otherwise was to restrict the bountiful hand of God to this one planet among all those in the universe. In doing this, Pope John XXI was actually trying to condemn the 219 propositions made by St. Thomas Aquinas regarding limitations on God's power in the interest of the powers of science. One of the condemned propositions stated, "That the first cause [i.e., God] could not make several worlds."

Pope John XXI's insistence that God could indeed create anything He wanted to led to questions of whether there were other worlds out there sustaining life.

35. The Magellanic Clouds were named for explorer Ferdinand Magellan. They are two galactic systems next to our galaxy and are visible to the unaided eye in the South Hemisphere. This detailed photo of the Large Magellanic Cloud, located in the constellation Dorado, was taken in 1897.　　　　　　　　　　　(Harvard College Observatory)

Chapter 4

THE RENAISSANCE OF MARS

The Renaissance of the fourteenth and fifteenth centuries was a time when the Greek forms of classical knowledge and art were being rediscovered, based on translations and excavations made during the Middle Ages. Greek and Roman manuscripts on science, philosophy, and mythological plays slowly filtered down from the Islamic world, and the first universities were founded in Europe to house these books, allowing scholars to study them.

The advent of the printing press in the mid-fifteenth century had a dramatic effect on Western learning. By 1500, approximately seven million books were in print of more than thirty-five thousand titles. This allowed scholars to easily consult the literature and science of the ancients without having to travel vast distances to universities.

Ptolemy's list of the stars was used by the German painter and printmaker, Albrecht Dürer, when he created the first printed star maps in 1515. His charts included both constellations and stars in a pair of beautiful planispheres.

The rapid development of both art and science became intertwined during the Renaissance. The construction of vast Gothic cathedrals had led to a better understanding of geometry, which was then used to develop the Renaissance perspective in the visual arts. As artists became more concerned with the relationship between art and observer, trying to create an illusion of the world, they studied nature, particularly the effects of light on objects and the proper anatomical proportions of people.

36. Leonardo da Vinci's sketch (ca. 1510) of a man in an "ornothopter" is only one of the fantastical flying machines that he hoped would lift man into the heavens. Da Vinci also invented a device that ground concave mirrors, resulting in a telescope by 1509, a century before Galileo's.
(Scala/Art Resource, NY)

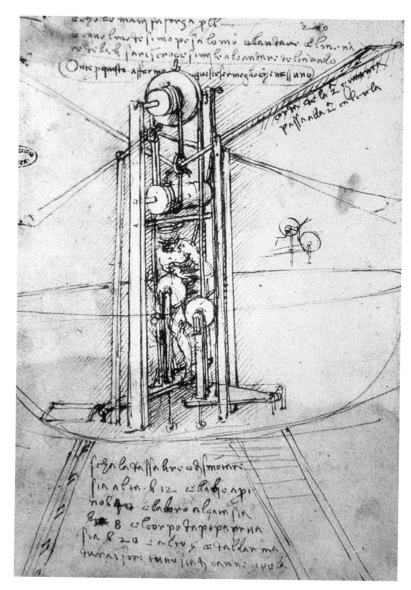

Leonardo da Vinci was brilliant as both an engineer and an artist. Da Vinci studied the flight of birds, the movement of water, geology, mechanics, military engineering, the growth of plants, human anatomy, among many other things. Yet for all his inventive spirit and scientific observations, da Vinci's researches were guided by his artistic temperament.

NICOLAUS COPERNICUS

Nicolaus Copernicus (1473–1543) was the first to transform the ancient models into something new—a model of the universe that reflected the actual movements seen in the sky. He put the Sun at the center of our solar system and set the planets into motion around it.

Copernicus based his theory, which he compared to Ptolemy's *Almagest*, on direct observation. Prior to the printing press, inaccuracies in Ptolemy's calculations were attributed to

37. Perugino was named for his hometown of Perugia, where he painted this early rondel of Mars (ca. 1470). Perugino was the teacher of Raphael, and though he helped to shape the High Renaissance, his figures hark back to the Middle Ages, created out of linear patterns rather than forms in three-dimensional space.

(Scala/Art Resource, NY)

transcription errors, but as the owner of a number of printed editions Copernicus could analyze the mathematics for himself.

He believed that a Sun-centered universe accorded better with the Platonic ideal of harmony, and it also fit his observations. Yet Copernicus did not publish his *De Revolutionibus* until 1543 when he was on his deathbed. It could be that he feared religious persecution. Though the Reformation was loosening the chains of Rome, even Martin Luther stated that Copernicus "wishes to reverse the entire science of astronomy; but sacred Scripture tells us that Joshua commanded the sun to stand still, and not the earth." Charges of heresy were taken seriously, and the fist of the Inquisition was greatly feared.

38. Nicolaus Copernicus was dissatisfied with Ptolemy's Earth-centered universe and displaced it with a heliocentric cosmology, with Earth orbiting a stationary sun. (Author)

Macula et Faculae ex uariis obferuandj modis, ftabiliuntur.

THE REFORMATION

The Reformation in the sixteenth century was fundamental in shattering the Church's hold on culture, allowing questions about existence to be more vigorously pursued. The Reformation was intended to be a religious reform, to return to the fundamental precepts of Christianity as stated in the Scriptures. But by the end of the seventeenth century, science had replaced Christianity as the focal point of European civilization. Instead of looking for the reasons behind material existence (something Platonic ideals and Christianity had in common), people began to ask practical questions of "how." Science was turning out to be more useful than theology and philosophy; it was testable, tangible, and inevitably generated even more insights and innovations with each experimentation.

The scientific revolution began, just as it did in ancient Babylonia, with astronomy. It is no wonder that humans continue to

look to the stars for signs of advancement; it is part of our heritage to believe that our future resides in the sky.

JOHANNES KEPLER

It was Johannes Kepler's study of the motion of Mars that finally led to the discovery that the orbits of the planets are elliptical, not circular as called for by the Ptolemaic model.

Just as the ancient Greeks believed that physical forms were a reflection of ideal perfection, Kepler (1571–1630) believed that God must have constructed the world based on principles of order and harmony. Kepler spent his life looking for simple mathematical relationships to describe the motions of the planets.

It was not originally Kepler's choice to study Mars. In 1600, Kepler went to Copenhagen to study with Tycho Brahe at his advanced observatory on an island in the Sund. This observatory included a chemical laboratory, printing plant, and quarters for visiting astronomers. Kepler brought a new depth of insight to Tycho's magnificent observations. But the relationship between the two men was stormy. Tycho rejected both the Ptolemaic and

40. Today, a geological feature on Mars is called Kepler's Ridge in honor of the mathematician's astute guess that Mars would have two moons. Yet he was wrong about the number of moons around the other planets, underestimating the abundance of moons in our solar system. (NASA/JPL)

41. Tycho Brahe was a Danish nobleman of Swedish extraction who devoted his life to astronomy. In his attempt to prove that Earth was the center of the universe, Brahe measured the positions of the planets to an unprecedented degree of accuracy, a major achievement considering that in the sixteenth century astronomers did not use telescopes. (U.S. Naval Observatory)

Copernican models and disdained Kepler's philosophies. To prove his theories that there were more bodies in the heavens than could be accounted for by a spherical system, Tycho studied the supernova in 1572, stating "that it is neither in the orbit of Saturn . . . nor in that of Mars, nor in that of any one of the other planets. . . ."

Surely it must have been spite that made Tycho assign Kepler the study of the movements of the most difficult of all planets—Mars. Kepler had to face the fact that the inaccuracies of the Ptolemaic model were never clearer than when examining Mars. This is partly due to the proximity of the fourth planet, which makes it easier for astronomers to precisely pinpoint its position in relation to the other stars.

Kepler believed that with the amount of Tycho's data at hand he would be able to determine the orbit of Mars in eight days. Yet a year later, when Tycho died, Kepler was still working on a theory to explain the motion of Mars.

It was not until 1609 that Kepler took the daring intellectual

42. This painting of the sky at night from the planet Mars, published in the French magazine *L'Illustration*, shows the view as Kepler must have imagined it, triangulating the planets and the stars in his mind.　　(Mary Evans Picture Library)

43. This engraving of a telescope was made in 1647. Thirty-five years earlier, using the same type of telescope, Galileo observed that the Moon was not smooth as Aristotle had claimed, but that it had jagged mountains and crevices. Galileo also charted the different phases of Mars and he observed satellites around Jupiter. The moons of Mars were too small for him to see. (Archive Photos)

leap of imagining himself on Mars, attempting to determine the path of the Earth's motion against the stars. When pages and pages of calculations did not reveal the answer, he tried to picture what the motion of Mars would look like from the position of the Sun. His answer was the ellipse, with the Sun positioned at one of its foci. And the orbit of Mars—like that of Mercury—is much more elliptical than the other planets', making this discovery an even greater achievement.

Kepler's discovery confirmed that the planets move faster when they are closest to the Sun and slower at the periphery of their orbits. This justified Kepler's belief in an ordered universe, and he compared the movements of the planets to musical harmonies. Just as the medieval chants were being replaced by polyphony, the music of many voices, Kepler was inspired by principles of ratios between the consonances that he found in Vincenzo's book *Dialogue of Ancient and Modern Music.* The variations in the speeds of the planets' orbits were exactly like a musical polyphony. In 1687, Sir Isaac Newton's *Principia* would provide the physical basis for Kepler's laws.

Kepler maintained such faith in the principles of harmony that when he heard Galileo had discovered the four moons of Jupiter he wrote in 1610: "I am so far from disbelieving the existence of the four circumjovial planets that I long for a telescope, to anticipate you, if possible, in discovering two around Mars, as the proportion seems to require. . . ."

Kepler had the worst luck when it came to working with other astronomers. He never got his longed-for telescope from Galileo, who replied that he had none to spare and it would be too difficult to craft a new one for Kepler. Rather than trust to Kepler's mathematical equations of elliptical orbits, Galileo supported Copernicus's theory of Sun-centered spheres.

Galileo Galilei certainly had nothing to fear from Kepler; he was the best-known astronomer of his time, a friend of Pope Urban VIII and backed by the Italian Academy of the Lynx (founded in 1603). Yet Galileo's shameless self-promotion partly contributed to his fall at the hands of the Inquisition. His 1632 Copernican manifesto, "Dialogue Concerning the Two Chief World Systems, Ptolemaic and Copernican," was banned. His biggest mistake was in naming the plodding defender of the Roman Catholic Church "Simplicio." Though Galileo attested that he damned and abjured his "errors and heresies," nothing could wipe out his cry of truth that "the Galaxy is nothing else but a mass of innumerable stars planted together in clusters."

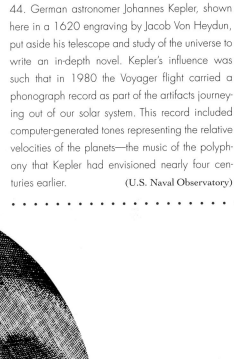

44. German astronomer Johannes Kepler, shown here in a 1620 engraving by Jacob Von Heydun, put aside his telescope and study of the universe to write an in-depth novel. Kepler's influence was such that in 1980 the Voyager flight carried a phonograph record as part of the artifacts journeying out of our solar system. This record included computer-generated tones representing the relative velocities of the planets—the music of the polyphony that Kepler had envisioned nearly four centuries earlier. (U.S. Naval Observatory)

As always, astronomers played the key role in the development of literature of cosmic perspective; Johannes Kepler's *Somnium* (*The Dream*; 1634) was developed from an essay intended to popularize the Copernican theory. In it Kepler described a dream journey to the Moon and the lunar mountains and valleys that Galileo had discovered with his telescope. He also envisioned inhabitants who had adapted to the extremes of hot and cold, constructing the hollows on the surface (a feat compared to Egyptian pyramid building and foreshadowing the great Martian canal theory). Kepler also described Earth as it would be seen from space. *Somnium* is considered one of the first works of science fiction.

Other early planetary tours included Athanasius Kircher's *Itinerarium Exstaticum* (*A Journey in Rapture*) in 1656. Kircher was a German priest and scientist who predicted the germ theory of disease, and his visionary round-trip visit to the planets was speculatively based on reality. Other writers, such as Bernard le Bovyer de Fontenelle in *Entretiens sur la pluralité des mondes habités* (*A Plurality of Worlds*; 1686), characterized inhabitants of other planets as gentle and ethereal, reminiscent of angelical beings.

THE ENLIGHTENMENT

Based on the Christian concept of linear, goal-determined time, intellectuals began to consider history as a record of human progress toward perfection. Francis Bacon located his New Atlantis in the future, helping to establish the concept of the Greeks and Romans as the young civilizations of the world while the current generation was heir to an accumulated reservoir of knowledge.

The Enlightenment's concern with nature, breaking it down through scientific observation, led to its objectification and a consequent rise of mysticism—as seen in the work of Emanuel Swedenborg who investigated a number of scientific fields, from

45. Il Guercino's realism relied on scientific principles of perspective and anatomy, yet his attention to details such as the shine on Mars's armor and the luminous skin of Venus is what makes this painting so arresting. In this 1620s composition, Il Guercino plays to the senses through vivid colors, luxurious textures, and the illusion of movement. The figures of Mars, Venus, and Cupid fill the space with dynamic intensity—embodying the height of the Baroque. Scala/Art Resource, NY)

• •

46. A mid-seventeenth-century painting of Mars by Diego Velázquez, the Spanish master. The Baroque style is characterized by contorted lines, asymmetrical rhythms, and stark contrasts of light and shadow.   (Scala/Art Resource, NY)

mathematics and linguistics to physics and geology. In 1743–45, Swedenborg underwent a painful religious crisis and began to focus exclusively on his own dreams.

In *The Earths in Our Solar System . . . with an Account of Their Inhabitants* (1758), Swedenborg described his visionary trip to the planets in our solar system, which was seen as having a spiritual significance. The book also contained some scientific speculation about the planets.

After his death, Swedenborg's followers founded the New Jerusalem Church to promote his doctrines. Swendenborg's writings influenced English poet William Blake (1757–1827) and German idealist philosopher Immanuel Kant (1724–1804) and were an important forerunner to the Romantic movement.

CHRISTIAN HUYGENS

In 1659, Christian Huygens began an exhaustive record of how the Martian surface markings moved with the rotation of the planet. He was the first to draw a surface feature on Mars, a windswept slope called Syrtis Major.

By observing the daily appearance and movements of the markings, Huygens determined that the rotation of Mars was twenty-seven hours. Only seven years later, in 1666, Gian Domenico Cassini of the Paris Observatory refined Huygen's rotational speed to a tighter figure of twenty-four hours and forty minutes—only two minutes and twenty seconds off.

These men were among the new intellects of the Enlightenment who believed that the environment created humanity. Christian Huygens, a founding member of the French Academy of Sciences in 1666, stated that God had made the worlds so they could be inhabited. Huygens went so far as to describe "planetarians" and their planetary environments in *Celestial Worlds Discover'd: Conjecture Concerning the Inhabitants, Plants and Productions of the Worlds in the Planets*, which was published before he died in 1690. Since general scientific opinion held that God could work only

47. This painting by Donato Creti shows early eighteenth-century observation in progress. The astronomer points low in the evening sky at the red star rising above the horizon: Mars. (Scala/Art Resource, NY)

within natural law, Huygens followed the principles of human nature and gave his planetarians two legs. He also gave them hands, which enabled them to write and acquire knowledge of mathematics.

With his astronomical observations, Huygens made a fairly accurate guess of the exact size of Mars, predicting that the diameter was about 60 percent that of Earth's (Mars is really 53 percent the size of Earth). He then measured the apparent size of the disk of Mars through a telescope and calculated the distance from Earth to the Sun as one hundred million miles—fairly close, as well, with an astronomical unit actually ninety-three million miles.

In 1672, an international expedition set out to observe Mars at its closest approach to Earth. Led by French astronomer Jean Richer, the group sailed to Cayenne, in South America, where their observations could be compared with those of the French Academy. By measuring the thousands of miles between the spots and the angle of their perspective, the distance was calculated by means of geometry. The result was the astronomical unit of eighty-seven million miles—six million miles short of the correct figure.

Still adhering to the Christian concept of time, even great scientists such as Johannes Kepler and Isaac Newton estimated that Creation dated to approximately 4000 B.C.

Isaac Newton was the last of the astronomers whose science was still rooted in alchemy, bibli-

48. This illustration from a French edition of Jonathan Swift's classic *Gulliver's Travels* shows Gulliver gazing up at the floating island of Laputa. In the fourth book, the carnal life of the humanoid Yahoos is compared to the horselike Houyhnhnms who lead a life of reason. In one fell swoop, Swift managed to satirize both the intellectuals and the romantics of his day.

(Mary Evans Picture Library)

cal prophecy, and the manuscripts of Babylon. Yet Newton's theories based on his observation of inertia and action-reaction experiments would lead the way for Einstein, who formulated his theory of relativity two centuries later.

JONATHAN SWIFT

The end of the age of Enlightenment was marked by a classic, satirical novel, *Gulliver's Travels*. In 1726, Jonathan Swift wrote the highly successful tale of Lemuel Gulliver's journey to Laputa, a marvelous island in the heavens. Like Lucian before him, Swift used the Laputa island to satirize his own society.

A true son of the Enlightenment, Swift blended realism with mysticism. Swift knew about Kepler's prediction that two moons orbited Mars, so his Laputa scientists were knowledgeable of the twin moons of Mars.

Swift's amazing descriptions captured a wide audience. His account of intelligent horses was one of many absurdities he added to "vex the world" with his seemingly true-to-life account. The straightforward style had the same air of sober reality that made Orson Welles's *War of the Worlds* so effective.

Chapter 5

LIFE ON MARS
IN THE INDUSTRIAL AGE

he technological advances during the nineteenth century
served to improve the quality of life. Even the most unscientific
layperson could see that progress was improving living conditions, society, and the understanding of the universe. Both hor-
ticulture and breeding were deliberately exploited to create
better produce and animals, while machines began to make
more machines until the production of all sorts of goods was
faster and increasingly efficient. Europe shifted from an agrar-
ian, handicraft economy to one dominated by manufacturing,
the foundation of modern society.

The cultural transition from the Enlightenment to the Industrial Age is perhaps best typified by Charles Darwin's *On the Origin of Species* (1859). Darwin combined advanced technological
observation along with the Enlightenment belief in mankind's
progression toward perfection, coming up with the concept that
every form of life is related to all others.

49. The Romantic movement flourished during the nineteenth century, as seen in this illustration of a floating Martian city by Paul Handy in *Letters from the Planets* (1890). The duty of the artist was to interpret the mysteries that lay beyond physical reality. Many romantic writers were drawn to the revelations of astronomy and science, which served as a basis for their flights of fancy.

(Mary Evans Picture Library)

50. This 1852 photograph is one of the best daguerreotypes in existence of the Moon. It was taken by J. A. Whipple using a fifteen-inch refractor telescope at the Harvard College Observatory in Cambridge, Massachusetts. On the basis of photographs such as this one, scientists believed that if there were no "Moon people," then there must be "Martians."

(Harvard College Observatory)

Astronomy also contributed to the belief that all things were fundamentally related to one another. In 1859, two German chemists, Robert Wilhelm Bunsen and Gustav Kirchhoff completed spectroscopic instruments, which proved that celestial materials in near and deep space held the very same elements as those on our own planet. This was seen as proof that there could be life on other planets.

Based on the concept of adaptation and evolution, writers began to envision alien life as a real possibility, not as simply a fictional way of examining the quirks of our own society.

One of the first writers to use biology to design an alien life-form was the French writer Camille Flammarion. In both his nonfiction work *Real and Imaginary Worlds* (1864) and his fictional *Lumen* (1887), Flammarion created humans and aliens who fit into an evolutionary view of the universe.

The Martian life-forms in Flammarion's novel *Urania* (1890) were used to prove the immortality of man's soul through reincarnation on other worlds. Though the plot of *Urania* was basically a romance, Flammarion also described a fairly realistic Martian environment.

It did not matter if the fictional subjects traveled through space in horse-drawn carriages or on an ambulatory moon, as in

James Cowan's novel *Daybreak: A Romance of an Old World* (1896). Cowan's depiction of parallel evolution on Mars, which had its own incarnation of Christ and Christianity, was considered fascinating because of the contrasts it drew between humanity and technology.

Just before the turn of the century, Mars became the focus of another interpretation of Darwin's theory of evolution in Ellsworth Douglass's 1899 novel *Pharaoh's Broker: Being the Very Remarkable Experiences in Another World of Isidor Werner (Written by Himself)*. In Douglass's parallel world on Mars, there is a Martian society similar to that of Egypt during the time of Joseph. Douglass's hero turns out to be capable of replacing Joseph and carrying out his seminal acts. After all, Isidor Werner was originally a grain broker from Chicago, and what is a mere shift in worlds to a product of evolution?

ASTRONOMICAL ADVANCES OF 1877

Before the mid-1800s, the Moon was the favorite destination for fictional voyages. Yet, with the advances in optics that produced larger lenses for telescopes, observers discovered that the Moon was a dead and sterile world. The craters were seen and recognized as evidence of ancient bombardment preserved in the dust. The Moon clearly lacked an atmosphere and livable temperatures, and no plants grew on the surface.

While the Moon was dead, the surface of Mars could barely be seen with the aid of nineteenth-century telescopes. The distinct red color seemed interspersed with blue-green patches, which were thought to be oceans or vegetation. The polar caps seemed to wax and wane with the seasons and were generally believed to be composed of snow and ice.

The tilt of Mars is comparable to Earth's, around twenty-four degrees. This produces extreme effects on the Martian seasons and its weather, creating a northern hemisphere with a springtime lasting for 199 days, a summer of 182 days, an autumn of 146 days, and a winter of 160 days.

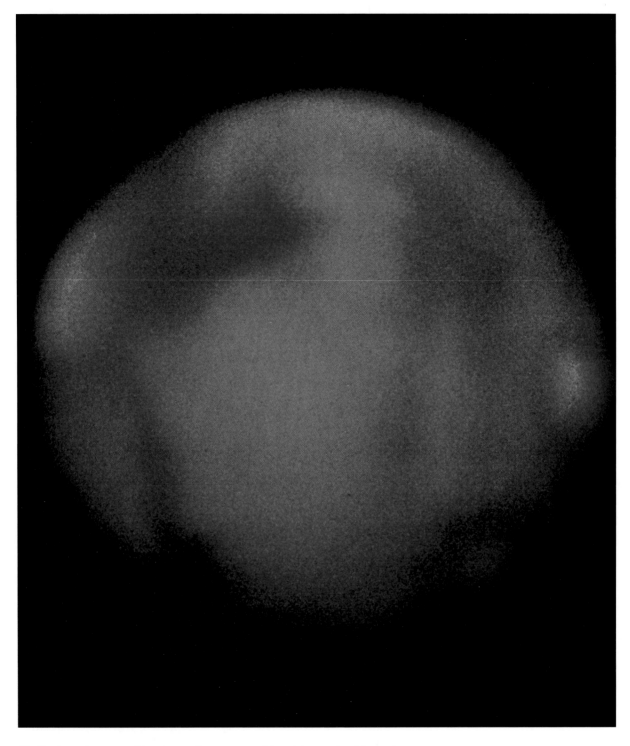

51. The best telescopes of the nineteenth century saw Mars only as a tiny, blurry orb. This photograph captures the distorting effect of Earth's atmospheric haze through the Catalina Observatory's sixty-one-inch telescope. Only the largest surface features can be detected, such as the planet's north polar cap.

(NASA/Lunar and Planetary Laboratory, University of Arizona)

52. This is the twenty-six-inch refracting telescope at the U.S. Naval Observatory in Washington, D.C., that astronomer Asaph Hall used to view the two moons of Mars. In Italy, young Giovanni Schiaparelli was seeing *canali* through a telescope one-third the size.   (U.S. Naval Observatory)

As such detail filtered down through trial and error, it was reasoned that if no one could live on the Moon, then Mars must be the place to look for life.

Astronomers were rewarded as they turned their eyes on Mars in the year 1877, when the orbits of Earth and Mars brought them into the best viewing position of the century. Observations made during this year by astronomers such as Asaph Hall and Giovanni Schiaparelli served to revive the myth of Mars as surely as the Roman Empire breathed new life into the Greek god Ares.

53. This twenty-second exposure of Mars shows the brighter moon, Phobos, is southwest of the planet, while Deimos is northeast and about twice the distance away. The dark shape around the planet is the shadow of the small metallic filter that was used to reduce the brightness of Mars.

(U.S. Naval Observatory)

Asaph Hall was one of the luckiest astronomers that year: he was able to conduct his research with the new twenty-six-inch telescope at the U.S. Naval Observatory in Washington. Within two weeks, Hall threw off the veil of confusion and proved the existence of two Martian moons. Hall was quickly joined by excited fellow astronomers who confirmed that Mars did indeed have two moons. Hall was given the honor of naming the moons, and he chose to preserve the mythological associations. Homer's *Iliad* described the sons and servants of Ares as Fear and Rout; Hall, honoring this tradition of Mars as a war god, called the moons Phobos and Deimos.

54, 55. Phobos; Deimos. As seen in these two photos, the Martian moons, Phobos and Deimos, are similar but not identical. Grooves and craters dominate the uniformly dark surface of Phobos at this resolution. Deimos, however, appears to be very smooth, with a few craters and areas of bright albedo. However, at higher resolutions developed during the late twentieth century, the surface of Deimos is revealed to be saturated with craters as well. (NASA/JPL)

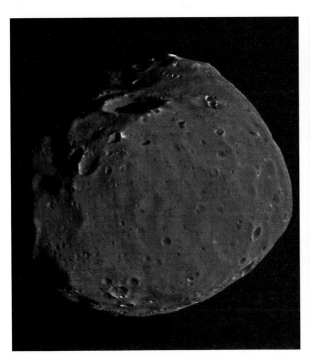

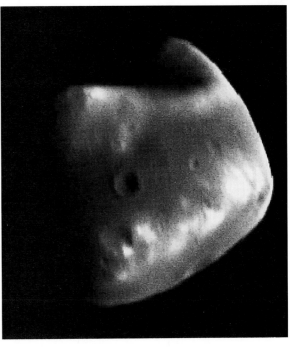

LIFE ON MARS IN THE INDUSTRIAL AGE

56. Mars is difficult at best to view through Earth-bound telescopes. Even this photograph captured by a Viking spacecraft as it raced toward the planet reveals indistinguishable markings of volcanoes, craters, canyons, rills, and patches of brightness from frost or ice.   (NASA/JPL)

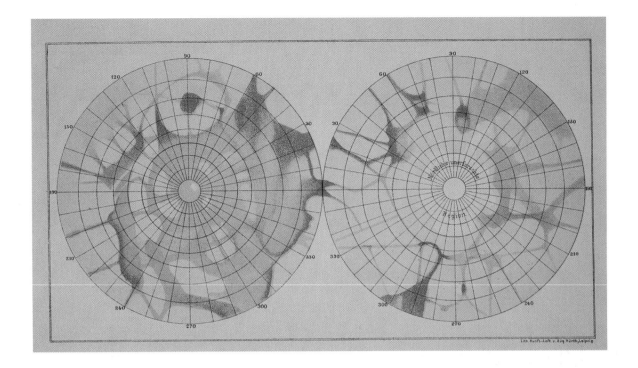

GIOVANNI VIRGINIO SCHIAPARELLI

Other astronomers who viewed Mars during the banner year 1877 did not receive such immediate and reassuring confirmation of their findings. Yet undoubtedly the most notable (and notorious) astronomer of the nineteenth century was Giovanni Virginio Schiaparelli. Schiaparelli created a lasting cultural impression of Mars and its potential inhabitants when he announced he had observed enormous straight lines lacing the surface of Mars. (Other astronomers had reported seeing lines or streaks on the surface of the red planet, but these were usually dismissed as vision quirks caused by optics.)

At first Schiaparelli doubted his own observations; Mars was difficult at best to view through an eight-inch telescope. But as he looked on successive nights, again and again he saw the prominent lines across the distant world. Schiaparelli released a record of his observations, noting that many of the lines emerged from the darker areas of Mars.

In his announcements Schiaparelli used one word, *canali*, which ignited intense interest in Mars on a level never known before. In Italian, *canali* means "channels" or "grooves." Yet

57. In 1879, Italian astronomer Giovanni Schiaparelli drew this map of the hemispheres, complete with the *canali* that he could discern through his telescope. This map was reprinted in a German science book.   (Mary Evans Picture Library)

both scientists and amateurs tended to translate this term into "canals," thereby implying that the lines were artificial waterways. And so, the reasoning went, such waterways could only be created by a highly intelligent race determined to conserve their diminishing water supplies on an old and decrepit planet.

Debunkers of the idea of canals protested that they would have to be thirty to fifty miles wide in order to be seen on Earth. Even the Nile was not wide enough to be seen from such distances. Yet later journeys into space proved that the Nile *could* be seen because of the wide band of vegetation that bordered it, making it appear as a straight band—exactly like Martian *canali.*

Nineteenth-century astronomers could not imagine that nearly a century later NASA astronomers would photograph the Valles Marineris canyon on Mars, the largest natural feature discovered so far in our solar system. It averages 250 miles across and stretches some 3,000 miles, the distance from Los Angeles to New York City.

For years Schiaparelli observed the planet Mars, drawing maps and charts that began to fill library shelves all the world over. In an era of astronomy, which relied on superb eyesight and attention to detail, he was able to draw the features that photographic plates could not record.

Schiaparelli even updated his maps, noting the double line of the Nilus "canal" and the color changes in the land masses. He believed these masses were giant swamps,

58. There are many possible explanations for what early astronomers believed to be canals on Mars, but closer examination would have to wait until the Viking missions of the 1970s. In this photograph of Mars, the natural channels extending into the Chryse Basin form troughlike structures that could have been detected by Schiaparelli. (NASA)

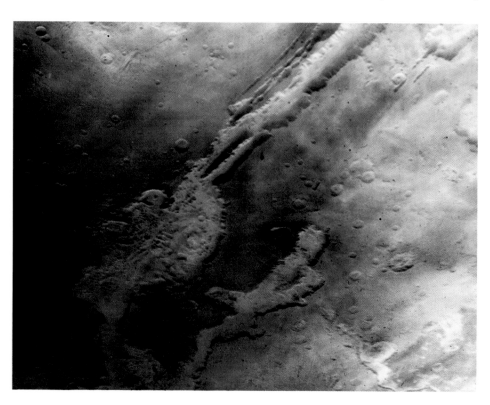

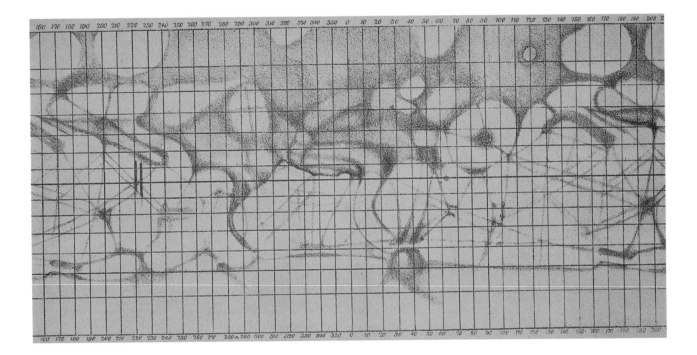

shifting from a dominant orange to yellow, brown, and black. The map of Mars soon became crisscrossed and blotched with an almost bewildering array of Schiaparelli's markings.

It was nearly eight years before confirmation of Schiaparelli's maps began to roll in from observatories around the world. Stanley Williams in England, H. C. Wilson in the United States, and Perrotin and Thollon at the Nice Observatory stated that they had witnessed and confirmed the markings displayed on Schiaparelli's maps. These confirmations would lure on a later generation of astronomers, like Percival Lowell who searched for canals for decades.

Schiaparelli confounded his supporters by viewing his own findings with extreme caution. While he insisted that the Martian features he mapped were authentic, he stressed that they might be natural rather than artificial structures. The problem, he repeated again and again, lay in the fact that Mars was enormously difficult to view—partly due to its fast rotation and to Earth's own atmospheric interference.

Despite the controversy and the difficulties of studying Mars, Schiaparelli continued to faithfully observe the red planet until his eyesight failed in 1892.

59. The appearance of Schiaparelli's maps, such as this sketch by the astronomer completed in 1881, inspired numerous maps drawn by both amateurs and professionals. Many people joined the craze of searching for the "canals" that were believed to be evidence of intelligent life on Mars. (Mary Evans Picture Library)

The observations of astronomers such as Schiaparelli, Williams, and Wilson encouraged a widespread conviction that there was indeed intelligent life on Mars. Schiaparelli even contributed to the final postscript of Louis Pope Gratacap's 1903 novel, *The Certainty of a Future Life on Mars*, which copied Flammarion's reincarnation theory as the basis of alien life on Mars.

The general belief in Martians became so pervasive that Mars was excluded from the Prix Guzman contest, which offered to reward anyone who established interplanetary communications with an alien society. Madame Guzman, who established the award and provided the prize money of one hundred thousand francs (then equal to about twenty thousand dollars), said that she excluded Mars because it would be too easy to establish contact with that planet.

Most astronomers adhered to the theory that a planet's distance from the Sun was also a measure of its age (which made sense if you believed in the concept of progressive evolution). Martians were usually imagined as socially superior to Earth's inhabitants, as in L. Edgar Welch's *Politics and Life in Mars: A Story of a Neighboring Planet* (1883). Welch's novel compared a

60. Armed with the new scientific conclusions gathered by Hall and Schiaparelli, artists and writers began to imagine the environment of Mars for newspapers, magazines, books, and even advertising such as this cigarette card.

(Mary Evans Picture Library)

WILLS'S CIGARETTES.

IMAGINARY LANDSCAPE ON MARS

61. The earliest fiction concerning Mars was based on eloquent fantasy rather than fact. These stories were similar to those about vampires, monsters, sorcerers, lost islands, and voyages to the center of the Earth or to the bottom of the sea. This illustration shows the Venusian spacecraft that visits Mars in a series of nine magazine stories, "Letters from the Planets" (from 1865) by W. S. Lach-Szyrma.　　　(Mary Evans Picture Library)

rather backward Earth to the socialist reforms developed by the Martian civilization. Even when Martians were seen as a lost race of earthlings, as in Hugh MacColl's *Mr. Stranger's Sealed Packet* (1889), the society was described as older and wiser.

Based on the evidence of the "canals," many people believed that Martians must possess greater technological prowess. Some people even suggested that huge light installations should be set up on Earth to aid advanced Martian observatories in studying our planet. Proposals of how to contact our neighbors flew thick and fast.

English writer Percy Greg required two volumes for his scientific explanations of a man's journey to Mars in *Across the Zodiac: The Story of a Wrecked Record* (1880). In this technical look into the future, Greg developed a force field that worked by "negative gravity," the precursor of antigravity space travel found in many later sci-fi novels. This solved the problem of sufficient fuel for the long trek to Mars—not to mention the energy and supplies needed for the return to Earth. Robert Cromie's novel *A Plunge into Space* (1890) also incorporated the use of antigravity to power a spheroid spacecraft that traveled to Mars.

Even more fascinating than Greg's technology was his description of the Martian civilization. The Martians had devel-

62. Lach-Szyrma's "Letters from the Planets" focus mainly on sightseeing and ethics, yet they also include the warning that other planetary environments could lead to vastly different customs. Paul Handy made these unbelievable tales come alive with his exquisite illustrations, such as this drawing of a Venusian spacecraft on a Martian canal (1890).   (Mary Evans Picture Library)

oped a form of utopia based on advanced technology and telepathic ability, which enabled them to punish people for wrong thoughts. Some Martians were not quite so advanced, and they considered women to be property. The hero from Earth became embroiled in a Martian civil war on the side of the telepaths who fought against slavery. After losing both his friend and his wife in the final

63. In 1880, Percy Greg's novel *Across the Zodiac* attempted to be technically correct. Greg's hero travels to Mars in a huge spaceship propelled by an antigravity device called "apergy." But readers were even more fascinated by Greg's description of Martians who were very similar to Earthlings.   (Drawing by James Akins)

conflict (though his side won the war), he escaped back to his spacecraft and returned to Earth.

INVASION BY MARTIANS

The widespread belief that there *must* be life on Mars inevitably led to questions of whether Martians would be friendly. Throughout the ages, scientific speculation has been linked to weapons of war, and new technology made more frightening ways to die at the hands of both real and imagined enemies.

When the German Empire was consolidated after the Franco-Prussian War of 1870, the strength of the new German Army prompted the rearmament of the British Army. The worst-case scenario was dramatized by Sir George Chesney in *The Battle of Dorking* (1871), a drama-documentary illustrating the ease with which an invading German Army might reach London. It caused an immediate sensation and initiated a debate that continued until World War I broke out.

The possibility of space travel became much more real when an obscure Russian teacher and inventor, Konstantin E. Tsiolkovsky, published chemical tables for fuels, a design, and a mathematical formula for a rocket that could be flown. Scientists gave serious credence to his 1896 work on aeronautics, *Exploration of Cosmic Space by Means of Reaction Devices.*

64. The Martians in Warwick Goble's illustrations of *War of the Worlds* were terrifying, able to snatch up humans and stalk across the terrain in their spindly-legged spacecraft. *Pearson's Magazine* is perhaps best known for having published H. G. Wells's classic invasion story.

(Mary Evans Picture Library)

The American Rocket Society grew steadily in numbers and influence, while England had its highly respected British Interplanetary Society. But nothing could match the power and financial strength of Germany's rocket programs, with extensive research funded by the government. Technological competition among these countries naturally led to the idea that the Martians might beat *us* to the punch and arrive on Earth before we made it to Mars.

One Martian who managed to arrive, by telepathic means, was Thomas Blot's *The Man from Mars: His Morals, Politics, and Re-*

65. Wells never managed to resolve the conflict between his dreams of utopia and his interpretation of Darwinian "natural law." The English, who were so quick to invade other countries in the name of progress and technological power, narrowly escape their own humiliating defeat at the hands of Wells's invading Martians. In this 1898 edition, the Martian machines are examined after the aliens die. **(Mary Evans Picture Library)**

ligion (1891). Blot's vision took a truthful twist in portraying the Earthman as a hermit: the only man to hear the traveler's tales of a Martian utopia kept the knowledge of aliens to himself.

Aliens had much more impact, according to the German writer Kurd Lasswitz, when a Martian space habitat arrived at Earth in *On Two Planets* (1897). From their vantage point above the North Pole, the Martians subdued the human race and imposed a benign rule that aided human advancement. This book included technological speculation on the potential of mankind and was deeply influential on several generations of German youth.

H. G. WELLS

H. G. Wells's *War of the Worlds* (1898) was a logical extension of the nineteenth-century future-war story. First Wells published a brief vision of Mars in "The Crystal Egg" (1897), then followed it up with his archetypal alien invasion story.

Wells's novel firmly planted in the popular imagination the image of Martians as monsters. Yet his intent was to satirize English society, so smug in its power to colonize and dominate the entire world. Wells made a sharp point of their cowardly behavior when faced with the same fate in *War of the Worlds*. It has also been said that Wells was drawing a direct parallel between the European invasion of Tasmania and the Martian invasion of Earth.

Wells's sci-fi works of the nineteenth century were labeled "scientific romances" by reviewers. Wells is considered a romantic because he always stuck to his dreams of utopia. Yet his biological training made him aware of the natural imperative of survival of the fittest, as *War of the Worlds* so clearly illustrates. It is the height of irony that the all-powerful aliens were destroyed by the lower forms of life on Earth.

EDISONADE

Soon after Wells's *War of the Worlds* appeared, *The New York Journal* commissioned Garrett P. Serviss to write an unofficial sequel. Serviss's story, "Edison's Conquest of Mars" (1898), can be considered as a rebuttal to the British Martian invasion story.

Serviss dismissed the fear of Martian superiority with true all-American confidence. His fictional Edison went on a rampage, invading Mars with his unique weaponry—a disintegrator and antigravity devices. The subsequent war melted the polar ice cap and drowned all the Martians.

This story line is typical of the genre known as "Edisonade," considered to be the precursor of modern-day science fiction. The hero is usually a young male, modeled on the real-life hero Thomas Alva Edison (1847–1931), inventor and improver of a remarkable number of gadgets from the light bulb to the phonograph. The fictional hero inevitably uses his ingenuity to invent devices that save himself, his friends, and his nation from alien oppressors.

Other popular Mars-based Edisonades were Weldon Cobb's *To Mars with Tesla* (1901) and *At War with Mars* (or *The Boys Who Won*; 1897), and Langston Moffett's *The Conquest of Mars*. Cobb's stories were based on Nikola Tesla (1856–1943), the inventor who discovered the rotating magnetic field, the basis of alternating current machinery. Cobb's young hero was called

66. Sci-fi stories of alien wars eventually slipped into a genre known as "space opera." Prototypes of the genre included Stanley Wood's *Stories of Other Worlds* (1899). In this illustration of an Earth spacecraft destroying a Martian vessel, you can find the fundamentals of space opera: the heroes are constantly thrown into interplanetary conflicts and they are not shy about butting in and setting things to rights, saving the universe in the process. (Mary Evans Picture Library)

Edison, while J. S. Barney in *L.P.M.: The End of the Great War* (1915) called his hero Edestone.

This basic Edisonade plot has appeared in drama throughout the twentieth century, illustrating the American preference for heroic overcompensation. Ronald Reagan would ultimately end up in the White House with his "Star Wars" program—all in the name of making the world safe for humanity.

67. This image of an interplanetary rocket is from a 1928 edition of *Le Petit Inventeur*, only one of the numerous invention and electrical magazines to spring up by demand of a curious public.
(Mary Evans Picture Library)

Chapter 6

MARS IN THE EARLY
TWENTIETH CENTURY

The all-time most devoted observer of Mars was unquestionably Percival Lowell (1855–1916). Lowell was convinced that Schiaparelli's canals were real, establishing concrete evidence of advanced intelligence at work on Mars.

In 1894 a huge observatory was built on top of a mountain near Flagstaff, Arizona. It was a private observatory and Percival Lowell was its astronomer. The telescopes peered into space through the pristine air covering the high deserts of the American Southwest, as Lowell searched for proof of alien life that he was certain existed on Mars.

Percival Lowell was popular with the general public, for he was unlike other scientists, especially because of his impressive air of high breeding. With his personal fortune, he did not have to prove anything to funding agencies or governments.

Lowell made Martian studies an acceptable scientific discipline. Despite his nonacademic background, he was slavish to

68. Lowell's telescope magnified Mars more than 600 times, and even then, all he could see was a fuzzy target. Photographs through the giant scope were difficult at best. That left him with hours of viewing, which resulted in colored sketches like this one. Compared to the digital mosaic photograph prepared nearly seventy years later, Lowell's drawings captured remarkable detail.

(Lowell Observatory)

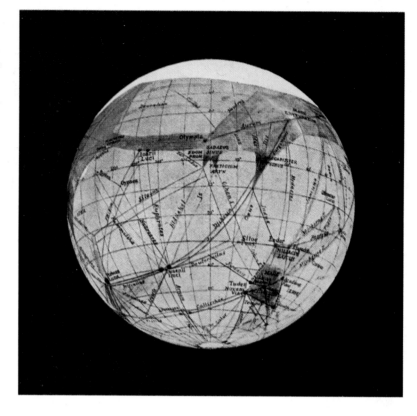

69. From 1896 to 1916, Percival Lowell spent most of his time at his twenty-four-inch refractor telescope at Lowell Observatory, studying Mars and mapping its surface features. This turn-of-the-century photograph records the astronomer's mission to keep the "proof of life on Mars" alive and well in the mythos of America.

(Lowell Observatory)

the truth as he knew and understood it. He even established hard data that the great bodies of water reported by Schiaparelli did not exist. Despite the fact, however, that his powerful new telescopes were directed on unflagging observation, proof of life on Mars was not found. Still, Lowell's enthusiasm remained undaunted.

Lowell was not the only observer frustrated by Mars. Between 1924 and 1926, Lacus Solis, a prominent Mars geographic feature, baffled observers by seeming to drift across the surface. Then several astronomers found Solis back where it belonged. No one could explain that mystery.

Prolonged observation of the tricky red planet commonly led to eyestrain, and most astronomers turned to cameras around the turn of the century. Yet even the finest photographic plates were so coarsely grained they could not capture the delicate lines of vegetation-laced canals fifty or even hundred of miles in width.

Yet Lowell insisted that the canals did exist. He believed the canals were a network of huge irrigation ditches, carrying water

70. This digital mosaic is one of the best images of the planet Mars that is available today. It was prepared by the U.S. Geological Survey in Flagstaff, Arizona, one of forty made from Viking photographs to give detailed, accurate images of the Martian surface to scientists who are preparing new missions to Mars.

(U.S. Geological Survey, Flagstaff)

from the melting polar caps to the inhabitants of Mars. He believed the light and dark areas reflected seasonal changes due to crop cultivation, as the older and wiser race attempted to revive an arid desert world. His theory was founded on the technological marvels of the Industrial Age: the completion of the 363-mile-long, 40-foot-wide Erie Canal connecting the Great Lakes with the Hudson River, and the 100-mile-long Suez Canal completed in 1869.

Lowell's 1896 book, entitled *Mars*, depicted a world uncannily like the desert of Arizona, colder than the Southwest due to the thin air, yet capable of sustaining life. Authors such as Mark Wicks, in *To Mars via the Moon: An Astronomical Story* (1911), fictionalized popular science, including Lowell's theories, for younger readers.

71. Lowell collected his data into series of book sets. This detailed globe-rendering of the red planet was completed by 1905. With evidence such as this, the public found it easy to believe that fascinating life-forms and cities had developed on Mars. (Lowell Observatory)

Lowell Observatory.   MARS—1905.

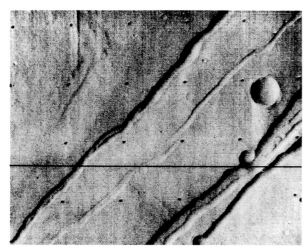

72, 73. These two Mariner 9 photographs reveal parallel features that early astronomers may have seen as canals on Mars. The left photo shows Martian rills, or cracks, that are part of a system of fissures that could have been caused by water freezing in the unknown past. The right photo includes part of the Vallis Marineris, a 3,100-mile "grand canyon" of Mars, with a maximum width of 62 miles.   (NASA/JPL)

SPACE OPERA

Lowell's enthusiastic view of Mars inspired a long, illustrious line of exotic, uninhibitedly romantic stories about Mars. This continued the fantastical, romantic tradition of the nineteenth-century novels of alien wars and utopian societies. Henry Wallace Dowding's *The Man from Mars, or Service, for Service's Sake* (1910) was a romantic meditation on the ideal civilization on Mars and its ornate ecology.

Edwin Lester Arnold created an even more lush environment in *Lt. Gulliver Jones—His Vacation* (*Gulliver of Mars*; 1905). The plot of Arnold's *Gulliver* was a thoroughly romantic takeoff of Jonathan Swift's *Gulliver's Travels*. Arnold really put a new spin on things by sending the hero all the way to Mars and having him rescue an exotic princess.

Telepathy became a favorite device in romantic science fiction. G. McLeod Winsor's first novel, *Station X* (1919), described a team from Earth and Venus fighting off a psychic invasion from Mars. John A. Mitchell's *Drowsy* (1917) was about a telepath who discovered antigravity and visited both the Moon and Mars. There was even a proto-Superman in "Vengeance of Mars" (in *Illustrated Chips*; 1912).

The desire for stories about Mars was in part filled by the early pulp magazines. The pulps were serialized novels done in the same style as dime novels with one story per issue. In the late nineteenth century, pulps such as the *Boys' Papers* helped create a solid readership for the future sci-fi pulp magazines and novels that emerged in the Golden Age of science fiction in the mid-twentieth century.

74. Many of the genre topics of what would become science fiction were developed in the early juvenile pulps. Even big-name authors such as Jules Verne appeared in British periodicals, with sixteen serializations in *The Boys Own Paper*. Fenton Ash first published *A Trip to Mars* (1909) two years earlier as "A King of Mars," in *The Sunday Circle*. (Mary Evans Picture Library)

Francis Henry Atkins, under the name Fenton Ash, published a large number of stories in the pre-sci-fi pulp magazines. Most of Ash's stories in *Boys' Papers* were lost-world adventures, but his writing was quite influential on the next generation of writers. Two of his stories are notable for anticipating Edgar Rice Burroughs's Barsoom novels by at least ten years: "A Son of the Stars" (in *Young England*; 1907–8) and *A Trip to Mars* (1909; in *The Sunday Circle* as "A King of Mars," 1907).

EDGAR RICE BURROUGHS

Edgar Rice Burroughs was inspired by Arnold's *Gulliver of Mars* as well as the romantic adventure stories found in the pulp magazines. Burroughs's Barsoom series was, in turn, much imitated by other authors, cementing the mythology of space opera. Space opera, in turn, would inspire the modern-day genre of sword and sorcery, exemplified by George Lucas's *Star Wars*, complete with light sabers, flowing cloaks, and battle armor.

Burroughs wrote numerous Barsoom books, mainly describing John Carter's battles against assorted villains and monsters on behalf of adoring Martian princesses. He was always armed with a sword, magic, and courage enough to face epic battles. Though the use of magic keeps Burroughs's stories from being true sci-fi, the imaginative Martian setting was enough to lure back generations of readers.

A Princess of Mars (1917; in *All-Story Magazine* as "Under the Moons of Mars," 1912) was the

75. This illustration by W. H. C. Groome (1909) shows a human "winging a Winged Martian." Ash's battling Martians foreshadowed Edgar Rice Burroughs's war-torn red planet by almost ten years. The wings on the Martians are reminiscent of angels, as is the saintly character of the Martian king. (Mary Evans Picture Library)

first Burroughs novel set on Mars. The Barsoom series followed different time constraints and physics, yet Burroughs based his Martian environment on the research done by Percival Lowell—as did everyone else in the early twentieth century.

Just as Lowell imagined, Burroughs's Barsoom civilization was dying in the harsh deserts of Mars. This caused constant strife among the various green-, yellow-, red-, and black-skinned people. John Carter fought beside great Martian warriors, and won the heart of Dejah Thoris, the red-skinned egg-laying princess of Helium. They lived together happily until Carter accidentally returned to the same cave on Earth where he had originally escaped Apache Indians by willing himself to Mars.

The personal saga of Carter was completed in *The Gods of Mars* (1918) and *The Warlord of Mars* (1919), both serialized in *All-Story* magazine in 1914. Eight other novels chronicled the adventures of Carter's family and friends, including: *Thuvia, Maid of Mars* (1920); *The Chessmen of Mars* (1922); *The Master Mind of Mars* (1928); *A Fighting Man of Mars* (1931); *Swords of Mars* (1936); *Synthetic Men of Mars* (1940); "Llana of Gathol" (in *Amazing Stories*, 1941); and "John Carter of Mars" (in *Amazing Stories*, 1941–43).

Burroughs's story lines involved moral dilemmas for his characters, and made certain that strength and fortitude were rewarded while weakness caused personal disaster. His heroes were inevitably heroic beyond their belief in their own powers, making the reader feel as if Mars were a dangerous and exciting place. Burroughs's ability to create believable characters can be seen in his other enduring fictional creation—Tarzan.

Along with H. G. Wells, Burroughs has had more imitators than other science fiction writers. One of the first imitators was also a competitor of Burroughs's—Otis Adlebert Kline, who began his Martian series with *The Swordsman of Mars* (1933). Other imitators, such as Lin Carter, Ralph Farley, and Gardner F. Fox, mostly re-created Burroughs's implausible fantasies, with varying success at establishing cohesive, imaginative worlds.

The Golden Generation of science fiction writers, including

Ray Bradbury and Leigh Brackett, owe a great debt to Burroughs. And the "rediscovery" of Burroughs during the 1960s inspired regular reprints of his books and a resurgence of interest in the space opera genre.

Kenneth Bulmer (under the pseudonym Alan Burt Akers) wrote more than thirty-five novels in the 1970s and 1980s that were Burroughs pastiches of fantasy–sword and sorcery. Terry Bisson was also inspired by Burroughs, as can be seen in *Voyage to the Red Planet* (1990), while Philip Jose Farmer's *A Feast Unknown* emphasized the more misogynist side of the Barsoom fantasies. In a story that was both satire and tribute to Burroughs, Farmer described the struggles of Lord Grandrith and Doc Caliban against the Nine, a secret society of immortals.

Despite decades of admiration and imitation of the Burroughs stories, there has been no official continuation of the Barsoom series.

76, 77. Even the most romantic works of art were based on contemporary scientific theories of Martian canals. Compare this illustration of the two faces of Mars by J. W. Potworthy (1927) with the color composites of the planet taken by the Hubble Space Telescope sixty years later. Where is the line between science and art?
(Mary Evans Picture Library and NASA)

MARTIAN MUSIC

Fiction was not the only arena to be inspired by the Barsoom tales: Sten Hanson's *The John Carter Song Book* (1979–85) was based on the information about Martian music found in Burroughs's novels. The vocals were created by computerized synthesizer to accompany the Martian music.

Mars was immortalized in music as early as 1920, Gustav Holst's *The Planets*. Like his contemporary Burroughs, Holst used a romantic style, reviving the choral tradition of folk songs and church hymns. Yet he was also influenced by Stravinsky's innovations and experimented with parallel harmonies, lopsided rhythms, and episodic, loosely structured forms.

The twelve segments of *The Planets* reflect the mythology of the Olympians. The first segment is Allegro, "Mars, the Bringer of

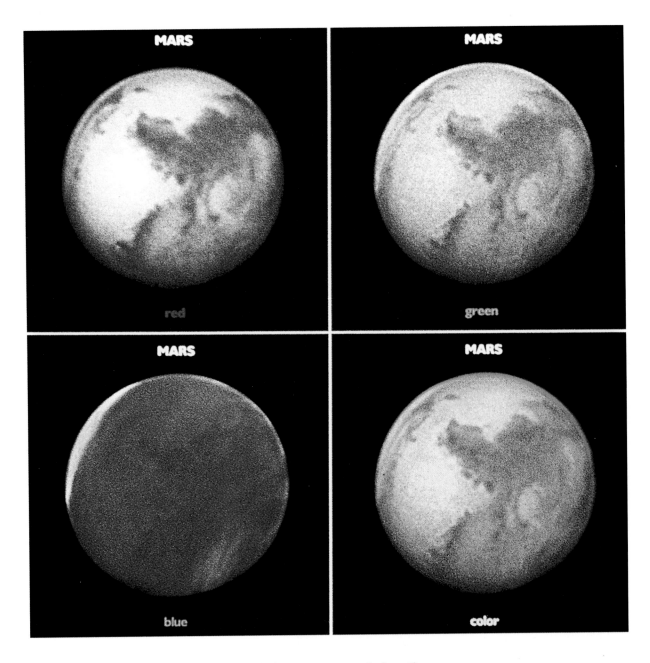

War." Holst's daughter Imogen agreed that Mars sounded as if it "had been commissioned as background music for a documentary film of a tank battle." But the tank and much of modern warfare had not yet been invented when Holst wrote it. Such is the power of music to catch the symbolic tone of a subject—in this case, the ferocity of Mars.

Albert Einstein's development of the special theory of relativity in 1905 created a conceptual leap in our view of the universe. Einstein took Newton's understanding of mass to the next step of dealing with quantum physics. Einstein's groundbreaking theory was based on work done by Michael Faraday (1791–1867). Faraday found that electricity and magnetism are conveyed through space by invisible lines of force, called fields, rather than the old-fashioned theory of space being filled with a substance called "ether." Einstein concluded that mass cannot accelerate to the velocity of light, which is the constant in which electromagnetic radiation or waves can be observed.

Though many of the stories and novels about Mars were blatantly fantastical—making the same leap as Percival Lowell from the hazy markings of "canals" to the existence of an advanced civilization—there were other authors who preferred a more realistic approach, adhering to the true potential of science.

Among those who preferred to debunk Lowell's romantic fantasy was Alfred Russel Wallace, Darwin's co-discoverer of evolution by natural selection. In 1907, Wallace stated in a review of one of Lowell's books that the astronomer's calculations of the average temperatures on Mars were wrong. The air was much

78. This 1907 illustration by W. Kranz of the supposed Martian landscape appeared in *Himmel und Erde*. German authors published surprisingly advanced stories based on scientific reality as they knew it. (Mary Evans Picture Library)

thinner than Lowell had estimated, placing the temperatures permanently below the freezing point of water.

Wallace also pointed out the impossibility of using canals, which were exposed to cloudless skies and relentless evaporation, to transport water across such vast distances. He concluded that there could not be intelligent life on Mars.

Yet even the most scientifically based stories tended to cross that line between hard science and good fiction. As Garrett P. Serviss, writer of proto-sci-fi novels such as *Edison's Conquest of Mars*, stated in "Curiosities of the Sky" (1909): "[Astronomy] is the most impressive when it transcends explanation. It is not the mathematics, but the wonder and mystery that seize upon the imagination."

HUGO GERNSBACK

The true banner carrier of nineteenth-century Edisonade was Hugo Gernsback (1884–1967). Gernsback revered modern technology, and his interests in electricity and radio were taken to imaginative extremes in his first magazine, *Modern Electrics* (1908), which was soon renamed *Electrical Experimenter*.

For this magazine, Gernsback wrote a series of apocryphal scientific adventures of Baron Munchausen, including: "Munchausen Departs for the Planet Mars" (1915); "Munchausen Lands on Mars" (1915); "Munchausen Is Taught Martian" (1915); "Thought Transmission on Mars" (1916); "Cities of Mars" (1916); "The Planets at Close Range" (1916); "Martian Amusements" (1916); "How the Martian Canals Are Built" (1916); and "Martian Atmosphere Plants" (1917). The series was reprinted in *Amazing Stories* in 1928.

Gernsback has often been called the "father of science fiction," particularly because he founded the sci-fi pulp magazine *Amazing Stories*. Today, the Science Fiction Achievement Awards are named the Hugos in his honor; he was given a special Hugo in 1960.

Though the Golden Age of Science Fiction is considered to have begun around 1938 (the same year Orson Welles made the seminal *War of the Worlds* broadcast), the flowering truly began with the emergence of science fiction pulp magazines in the late 1920s.

The flourishing of science fiction cannot be separated from the intense interest in Mars that could be seen after World War I in exhibitions, such as the one at the Treptow Museum in 1927—the same year Charles Lindbergh made his nonstop solo flight from Long Island to Paris. The Treptow Museum, just outside Berlin, set up a "Mars Exhibit" that included maps of the planet, dozens of huge photographs, and an enormous collection of books and periodicals all relating to Mars.

Also, the technological means to finally cross the seemingly unlimited space between Earth and Mars came with Robert Goddard's experiments with liquid-propelled rockets in the twenties. In 1935, he became the first person to shoot s liquid-fueled rocket faster than the speed of sound.

As the Nazis increasingly made Europe hazardous for scientists, philosophers, and artists, people began emigrating to the United States. The resulting concentration of intellectual power was electrifying during the thirties. It was here that Einstein first heard Georges Lemaître's theory of the expanding universe, based on Vesto Slipher's observations of redshifts in nebulae, indicating that things in the universe were flying away from each other. With Einstein's enthusiastic support, Lemaître's theory quickly resulted in the modern-day cosmology of the big bang theory.

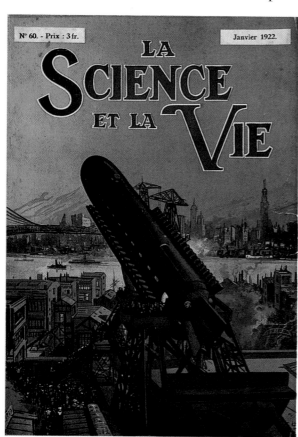

79. Robert Goddard's moon-rocket project was written up in this 1922 issue of *La Science et la Vie*. Goddard's scientific work was often popularized for a highly interested leadership of the early pulp magazines, bringing science and astronomy into homes around the world.

(Mary Evans Picture Library)

80. This photograph of a rocket launch was made on August 26, 1937, in the desert near Roswell, New Mexico. The numerous successful catapult launchings made by Robert Goddard heralded the dawn of the space age as surely as the Wright brothers' Kitty Hawk flights ushered in the air age.   (NASA, courtesy of Mrs. Esther C. Goddard)

81. The first proto–science fiction pulps tended to focus on occult fiction, such as *Thrill Book* (1919) and *Weird Tales* (1923). Then Hugo Gernsbeck published *Amazing Stories*, the first sci-fi pulp magazine featuring stories such as "Thia of the Drylands," by Hal Vincent, seen below in the cover illustration of the July 1932 issue.

(Mary Evans Picture Library)

82. If Hugo Gernsbeck is considered to be the father of science fiction, then his protégé, artist Frank R. Paul, would win the prize for father of science fiction illustration. Paul painted the illustration on the opposite page for *Fantastic Adventures* in 1939, as well as more than 175 covers for the sci-fi pulps. His aliens were favorites with the public, but his colors were considered to be garish due to the cheap three-color printing process that was used.   (Mary Evans Picture Library)

There was a tremendous reaction from the public in response to these technological developments. The demand for more science fiction stories was in a large part met by pulp magazines—easily produced, large tabloids that were printed on cheap paper. *Amazing Stories* (started in 1926) was the first pulp devoted exclusively to science fiction.

Most of the pulp writers were quite young and they let their imaginations run wild. There was such a large response from the public that the period has been called "revolutionary" by sci-fi critics.

The rapid turnover of stories that filled the pulps—*Argosy, All-Story,* and *Weird Tales*—led to a rough narrative style that emphasized real action at the expense of complex plots and believable characters. Gernsback unfortunately tended to emphasize the science over the fiction aspects, making the stories he included in his magazines rather dry.

Other early sci-fi pulps thrived on space opera, with plots echoing back to Edisonades and *War of the Worlds*. The first issue of *Amazing Stories* reprinted Edmond Hamilton's "Monsters of Mars" (1931), a quintessential space opera in which a man sets off with his companions (just like Jason and the Argonauts) to fight the threatening aliens.

As the pulps matured, they began to debunk the cliché of Martian monsters attacking Earth, as in P. Schuyler Miller's story about mistreated Martians in "The Forgotten Man of Space" (1933). Miller's other story about Mars, "The Titan" (part one, 1936), had a mild sexual content that was unacceptable in the moral climate of the times, and *Marvel Tales* ceased publication before the last installment.

Raymond Z. Gallun's Martians were also sympathetic in his series: "Old Faithful" (1934); "The Son of Old Faithful" (1935); and "Child of the Stars" (1936). Gallun's writing style was typical of the early thirties—bold and rough while avoiding sexual topics.

Like Raymond Z. Gallun, Jack Williamson, who wrote well into his seventies, had a tellingly space opera style. His *Beachhead* (1992) depicted an expedition to Mars that incorporated the knowledge gained by modern NASA probes, yet the plot echoed the swashbuckling of the pulps.

The most notable pulps evoked a sense of realism in their accounts of Martian life, as in Clark Ashton Smith's "The Vaults of Yoh-Vombis" (1932); C. L. Moore's "Shambleau" (1933); and Clifford D. Simak's "The Hermit of Mars" (1939). The turning point in the sophistication of Martian tales is often considered to be "A Martian Odyssey" (in *Wonder Stories*, 1934) by Stanley G. Weinbaum. Weinbaum's adventure story was set in a believable Mars environment, based on his understanding of chemical engineering.

One magazine stood out from the rest, *Astounding Science Fiction*, particularly after 1937 when John W. Campbell Jr. joined the editorial staff. Campbell guided the shift from space opera to a vision of Mars that rested on truth rather than fantasy. This was exemplified by P. Schuyler Miller's "The Cave" (1944), an ironic story in which an Earthman violates a sacred truce and is killed by Martians who must follow the customary laws in order to survive the long, cold night on Mars.

83. Jack Williamson believed that the early pulp magazines had an explosive effect on young audiences, especially those who grew up outside the cities. In his work he attempted to curb his "tendencies toward wild melodrama and purple adjectives" to achieve a more realistic science fiction story. This 1939 cover of *Astounding Science Fiction* features an illustration of the crash on Mars described in "Crucible of Power" by Williamson. (Mary Evans Picture Library)

84. The hard science of the 1920s and 1930s marked a shift away from fantasy depictions of space travel, such as William M. Timlin's *The Ship That Sailed to Mars* (1923). Timlin's illustrations were surrealistic and not as accessible to the masses—especially when compared to this crisp illustration of a 2038 spaceship, which includes loving attention to the most minute details of preparing a ship for a voyage to Mars (*Amazing Stories*, December 1938).

(Mary Evans Picture Library)

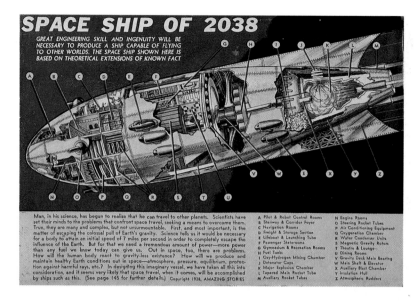

Campbell's list of authors while he was editor of *Astounding Science Fiction* reads like a who's who of science fiction: Jack Williamson, C. L. Moore, Lester Del Rey, Robert Heinlein, and Isaac Asimov. In the late forties, James Blish, Ray Bradbury, Arthur C. Clarke, C. M. Kornbluth, and Frederik Pohl were among the writers often published by the magazine.

COMICS

With Mars so visible in the pulp magazines, it was natural for the red planet to be picked up by another medium—comics. Like the pulps, the proto-sci-fi comics emphasized humor and fantasy rather than science, as in "Our Office Boy's Fairy Tales" (in *The Funny Wonder*, 1895), which was an anonymous British series about a family living in terrible conditions on Mars.

Sci-fi comics truly came into their own in the late twenties when Buck Rogers hit the international publishing world like a thunderbolt, complete with flying belts, antigravity metals, super propulsion systems, ray guns, and inventions that could reach beyond the boundaries of our solar system. The action-packed adventures in *Buck Rogers in the Twenty-fifth Century* were geared toward adults, yet the plotting was archetypal space opera. Buck, a lieutenant in the USAF, was inadvertently transported five hundred years into the future, where he found America overrun by hordes of "Red Mongols." Buck battled his nemesis, Killer Kane, with the help of his girlfriend, Wilma Deering. The Princess Alura of Mars and Wilma's young brother, Buddy, accompanied Buck Rogers in the Sunday version, which was much better drawn than in the daily papers. Buck Rogers comic strips were carried in both daily and Sunday newspapers from 1929 until 1967, spawning a radio show, movies, comic books, television shows, and much more.

The burst of enthusiasm for comics during the 1930s was partly due to the economic depression, when the public demanded cheap entertainment. They were willing to pay for visual formats such as movies and illustrated magazines, which served as a contrast to the all-pervasive power of radio.

The success of Buck Rogers sparked several rivals, among them *Flash Gordon* (1934 debut). The Flash Gordon comics were even better than Buck Rogers comics in quality—both in the drawing and the story lines—which in turn influenced the next generation of comics.

Flash Gordon appeared at first in Sunday newspapers, then later in the dailies. The series was even more reliant on the romantic elaborations of space opera, disdaining plausibility in its depiction of future technology and wallowing in swordplay and vicious monsters.

These comics were based on the mythology of classic literature, on the battles in the *Iliad* and the heroic adventures in the *Odyssey*. The plots were merely given a modern twist by placing the action in the future, often on Mars, where alien technology could make anything possible.

85. The *Flash Gordon* comic strip led to a popular radio serial, a short-lived pulp magazine, and in the late 1930s several film serials starring Buster Crabbe. The second film was *Flash Gordon's Trip to Mars* (1938), in which the setting is changed from Mongo to Mars.   (Everett Collection)

Science fiction cinema in the silent period was surprisingly sophisticated. Unlike modern sci-fi movies, which rely on ultra-realism for their effect, the silent films originated out of theatrical Expressionism. Two important early sci-fi films were *A Message from Mars* (1913) and the Russian-made *Aelita* (1924).

Mabel Winfred Knowles (under the pseudonym Lester Lurgan) wrote *A Message from Mars* in 1912, based on an 1899 play by Richard Ganthony. The silent film version of *A Message from Mars* was a gently ironic look at how a Martian's sensible advice affected a man here on Earth. The Martian's redemptive power was similar to that of the three ghosts in Charles Dickens's *A Christmas Carol* (1843).

Aelita, by Alexei Tolstoy, was also a satiric take on Mars. A group of Soviet astronauts traveled to the red planet only to find Martians living under an oppressive government. The astronauts incited a revolution, but the high point seemed to be when one of the astronauts taught the daughter of a Martian leader how to kiss.

These silent films tended to rely on their futuristic, Expressionistic sets rather than on complex, scientific explanations of how and why. This in turn influenced the design of the 1930s film serials of Flash Gordon, which were high on visuals and comparatively low on content.

86. Spring on Mars is captured in this Hubble Space Telescope view of the planet. The colors of Mars changed with the seasons, suggesting that vegetation was growing and dying on that distant planet. No wonder early-twentieth-century mankind looked to Mars as the likely home of an advanced civilization.   (NASA)

Chapter 7

THE GOLDEN AGE OF MARS:
POST—WORLD WAR II

Following in the tradition of Percival Lowell, Edison Pettit of the Mount Wilson and Palomar observatories spent years trying to confirm the existence of Martian canals. Pettit scanned Mars in several oppositions in the 1920s; when nothing turned up, he became a vocal disbeliever.

Then, on the morning of July 10, 1939, viewing conditions between the two planets were superb, with the sighting zone on Mars as close to perfection as Pettit had ever seen. As Pettit studied the mottled surface of Mars, a canal suddenly flashed into view. Before the morning passed he saw four canals, all visible at the same time.

In the following weeks, using a six-inch and a twenty-inch scope at Mount Wilson and working without interference or distraction, Pettit recorded clear observations of forty canals. He prepared charts and records, then announced his observations to other astronomers. Confirmation was received from Philip Fox,

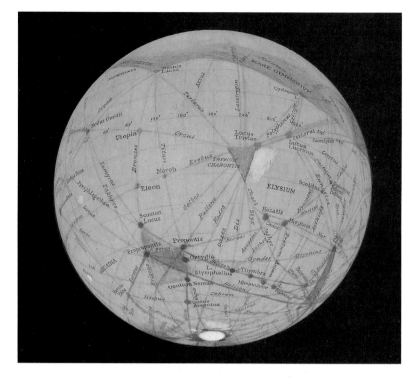

87. Despite the difficulties of observation, Mount Wilson Observatory's Edison Pettit and Lowell Observatory's Phillip Fox spent years photographing and sketching what they reported to be forty clearly observed canals on Mars. The two astronomers were using twenty- and twenty-four-inch scopes when Fox created this diagram.

(Lowell Observatory)

who used a twenty-four-inch telescope at Lowell Observatory to observe Mars during the same period in 1939. Fox agreed that there were indeed canals on Mars.

However, astronomer Robert S. Richardson, also at Mount Wilson Observatory, spent many disappointing evenings viewing Mars. Though he saw patches of color, he could detect no canals. The telescopes of the early 1940s simply were not precise enough to give astronomers a clear view of Mars.

Though most people were aware that surface conditions on the red planet were not conducive to life, the myth was strong enough to convince many scientists that there once *had* been life on Mars—a civilization great enough to construct canals, now ancient and crumbling. It was this same gut feeling that had led to panic on hearing Welles's broadcast about a Martian invasion.

Less than two months after Richardson's telescopic forays in 1941, all concern with the red planet evaporated in the thunder and fury of World War II. Yet the seductive power of Mars was so strong that Richardson made another attempt in 1954, viewing Mars through his six-inch telescope. Richardson claimed that *finally* he "saw the canal Indus, and probably Hiddekel and Gehon. . . ."

However, Richardson, Pettit, and Fox were all mistaken in what they saw on the surface of Mars. Thereafter, most telescopes of the world would ignore our neighboring planet, leaping far out into the void with its breathtaking secrets to be revealed and captured.

The Hale Telescope—a reflecting telescope with a 200-inch mirror—installed at Palomar Observatory in 1947 was easily the most powerful device of its kind in the world. It opened the doors to the universe, yet whenever its lens was trained on Mars, only a blurry orb could be seen. Ironically, the telescope was too powerful to view Mars.

88. Astronomers did not enjoy photographs of reasonable clarity of Mars like these two black-and-white shots until the 1940s. These views were taken with Palomar's 200-inch Hale Telescope, and, as you can see, the Hale scope is so powerful that a heavenly body as close as Mars is blurred. (Palomar/Caltech)

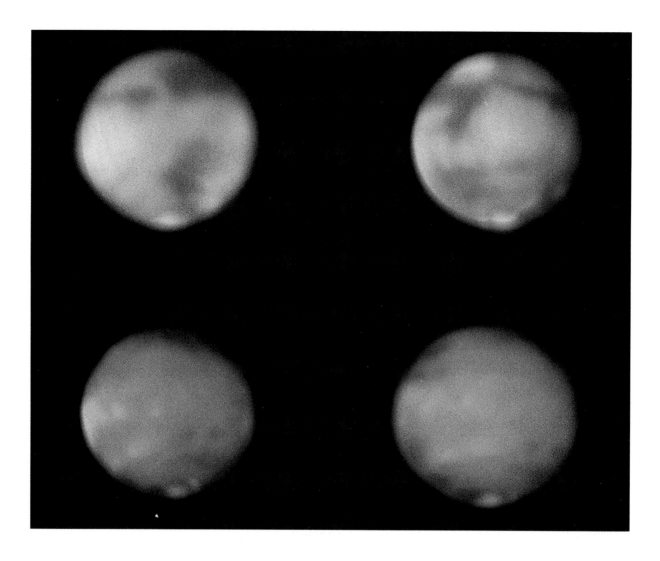

89. These color photographs were not taken by the Catalina Observatory's sixty-one-inch telescope until 1967. The comparison between photos of the 1940s and the 1960s reveals the slow progress that was made in Earth-based viewing of Mars. Little additional research and observation could be done until an orbital probe closed the gap between the planets.

(NASA/Lunar and Planetary Laboratory, University of Arizona)

. . . .

90. Mysterious Mars. In this dramatic color photograph taken as Viking approached the red planet, we can see the prominent slash marks of Vallis Marineris stretching across nearly a third of the planet. At the bottom is an extensive area of frost, and at the top a large white area of clouds can be seen.   (NASA/JPL)

. . . .

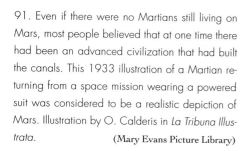

91. Even if there were no Martians still living on Mars, most people believed that at one time there had been an advanced civilization that had built the canals. This 1933 illustration of a Martian returning from a space mission wearing a powered suit was considered to be a realistic depiction of Mars. Illustration by O. Calderis in *La Tribuna Illustrata*. (Mary Evans Picture Library)

FANTASY—SCIENCE FICTION

Where the astronomers left off, fiction took over in sustaining the myth of life on Mars. The belief that an advanced Martian civilization had once flourished and perhaps had already died along with its planet inevitably led to questions of the survival of mankind. The invention of nuclear weapons and the world's involvement in the recent war prompted sci-fi writers to seriously examine our hopes for a future.

The stories about Mars during the post–World War II era were typically nostalgic and seductive in their elements of dark fantasy. Ray Bradbury's "The Exiles" (1949 as "The Mad Wizards of Mars") depicted literary and mythic characters exiled on Mars perishing when the last of the books containing their stories were burned or lost. As the emerald city of Oz disappeared, the Earth expedition on Mars was left in a vast, barren desert.

Ray Bradbury was a seminal figure when it came to the making of the modern myth of Mars. His *Martian Chronicles* (1950) defied science by presenting a Martian ecology that was conducive to human life. Other romantic authors also focused on biosystems that protected their sympathetic humans against all dangers, such as Robert A. Heinlein in *Red Planet* (1949), Cyril Judd in *Outpost Mars* (1952), and Leigh Brackett in a series for *Astounding Science Fiction* beginning with "Martian Quest" (collected in *Planet Stories*, 1940).

Leigh Brackett's swashbuckling style continued the space opera tradition that Burroughs had immortalized in his Barsoom Mars stories. Brackett emphasized the exotic qualities of the alien environment set in the distant past when humans and humanoid aliens interacted with one another. Humans were often portrayed as intruders on the decadent, dying Martian civilization, as in *Shadow over Mars* (1944) and *The Sword of Rhiannon* (1953; "Sea-Kings of Mars," 1949).

RAY BRADBURY

Bradbury's early work tended to be more fantasy than science-oriented fiction—perhaps even leaning toward the style of space opera. Yet some of the stories in the *Martian Chronicles* were published in mainstream magazines as well as in the sci-fi pulps, which catered to planetary romances, such as *Planet Stories*.

Though Bradbury bowed to common scientific knowledge in creating a Martian civilization that was long dead, the aliens, in their shape-changing forms, continued to haunt the human colonists. Bradbury's obvious yearning for more simplistic times and settings (such as midwestern towns) and the protagonists' quests for survival, in the epic sense, lent a rather somber tone to his work. Ultimately, however, the inhabitants were redeemed when the children, gazing into the water of the canal to see Martians, found a reflection of themselves.

Bradbury's stories were both whimsical and pensive, and they had a deep effect on other writers, even those who were concerned with a more realistic depiction of the red planet, such as Arthur C. Clarke, James Blish, and Ian McDonald. All owe a debt to Bradbury's resounding, romantic image of Mars.

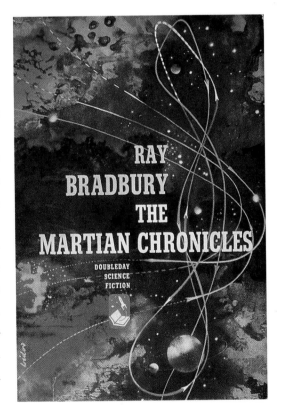

92. In *The Martian Chronicles* (1950), Ray Bradbury created a haunting version of man's colonization of the red planet. Bradbury took the myth of Martians to new heights, describing an advanced civilization that died out long before man ever dreamed of going to the stars, lending a modern plausibility to their ancient existence.

(From *The Martian Chronicles* [jacket cover] by Ray Bradbury. Used by permission of Doubleday, a division of Bantam Doubleday Dell Publishing Group, Inc.)

93. In 1980, *The Martian Chronicles* was made into a fairly effective TV miniseries, starring Rock Hudson as the spaceship captain. The haunting strength of the stories and the characters longing for survival was typical of the post–World War II era. Ray Bradbury transcended the boundary between science fiction and mainstream literature, and from there into popular culture.

(Everett Collection)

The demand continued to grow for realistic science fiction stories about Mars. In the spirit of the booming fifties, readers wanted to see humans move in and colonize the planet. No air or water? No problem. Hadn't the Allies just conquered the Axis armies? Wasn't the American economy soaring, with more babies being born into better lives than their parents had dreamed possible? Weren't near-miraculous household inventions appearing every day to make everything easier?

The public wanted realism when it came to Mars, and they got it in series of juvenile books, sci-fi pulps, novels, and movies. While the fantasies of the day tended to be thoughtful, the realistic sci-fi tales often had protagonists who liked to "get in there and get dirty," solving all problems from building space habitats to terraforming an entire planet.

Arthur C. Clarke's memorable story of humans terraforming Mars stood out from among the realistic depictions. His

94. The cheerful space opera of movie serials such as *Buck Rogers in the Twenty-fifth Century* and *Flash Gordon* gave way to the weak comic science fiction of the late 1940s, as in Bud Abbott and Lou Costello's *Invisible Man* movie and their *Abbott and Costello Go to Mars* (1953). Luckily for Mars, the sci-fi movie boom of the 1950s would reveal a whole other side of the red planet. (Everett Collection)

THE GOLDEN AGE OF MARS: POST-WORLD WAR II

view was generally optimistic, with a sustaining philosophy that technology could tackle any problem. Clarke's most famous work, *2001: A Space Odyssey* (1968), with Stanley Kubrick, was the culmination of Clarke's fictional longing for a superior race that could show earthlings the way—if they could only find the right way to ask. Despite intensely realistic prose, there was an underlying mystical tone running through Clarke's work.

Another sci-fi story that combined realism and a religious quest was *Conquest of Space*, screenplay by James O'Hanlon and produced by George Pal. This movie was remotely based on *The Mars Project* (1953), by Wernher von Braun, which was a solid sci-fi story. But the script of a military mission to Mars relied

95. George Pal produced *Destination Moon*, which was so successful that it initiated the sci-fi movie boom of the 1950s. Pal immediately followed his hit with *When Worlds Collide* (1951) and *War of the Worlds* (1953), directed by Byron Haskin. *War of the Worlds* was so popular that it was dubbed into foreign languages almost immediately upon release. (Everett Collection)

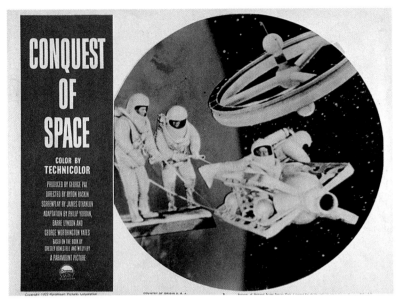

CONQUEST OF SPACE

COLOR BY TECHNICOLOR

PRODUCED BY GEORGE PAL
DIRECTED BY BYRON HASKIN
SCREENPLAY BY JAMES O'HANLON
ADAPTATION BY PHILIP YORDAN,
BARRE LYNDON AND
GEORGE WORTHINGTON YATES
BASED ON THE BOOK BY
CHESLEY BONESTELL AND WILLY LEY
A PARAMOUNT PICTURE

rather heavily on the religious fanaticism and the ultimate insanity of the commander for its plot twists.

The title for this film was taken from the popular-science book *The Conquest of Space* (1949), by Willy Ley and illustrated by Chesley Bonestell. Chesley Bonestell also did the matte work for the movie *Conquest of Space*, but the clumsy effects were nothing like the beautiful, photo-realist illustrations he painted for Ley's book. In a final, strange connection, Ley also wrote *The Exploration of Mars* (1956) with Wernher von Braun.

COLD WAR ANXIETY

A pessimistic worldview rose not long after the fury of military might—especially the bombings of Hiroshima and Nagasaki—had ended World War II. The dark specter of nuclear holocaust dominated the minds and imaginations of the public, and sci-fi writers responded with an ever-increasing number of accounts of alien wars.

Despite the rollicking of space opera battles in the pulps, a sense of pessimism invaded science-fiction during this time. The movie *Rocketship X-M* (1950) was rushed out to capitalize on the publicity generated by the first of George Pal's sci-fi films,

Destination Moon. In *Rocketship X-M*, astronauts were forced to land on Mars, where they discovered a race of mutants. But in the end the spacecraft crashed, killing the astronauts before they could warn mankind that the Martians had destroyed their own world in an atomic war.

Pelham Groom, a British writer, also explored the implications of nuclear war in *The Purple Twilight* (1948). His protagonist traveled to Mars in search of the lost island of Atlantis, and found instead a dying race of Martians who were unable to have babies because of the effects of a nuclear civil war. When the hero tried to return to Earth to warn mankind of the consequences of nuclear war, no one would listen.

One Cold War movie that truly flopped, yet serves as a historical artifact of McCarthy-era politics, was *Red Planet Mars* (1952). The premise was mildly interesting: a husband-wife

97, 98. In the film version of *War of the Worlds*, the walking tripods of the Martian craft are transformed into much more believable flying saucers. Yet the power of the classic story can still be seen in the 1953 poster below, with the clawed hand swooping down to grab the hapless humans, an almost identical reproduction of Henri Lamos's illustration in the 1898 French version of *War of the Worlds* in *Je sais tout.*

(Everett Collection)
(Mary Evans Picture Library)

science team received TV transmissions from Mars. One group of transmissions had been faked by an ex-Nazi in order to incite panic in the United States, while the other group had truly been sent from Martians proclaiming that they had direct contact with God. Of course, in the end, the godless communists were vanquished.

MONSTER MOVIES

The sci-fi movies of the 1950s are best known for their monsters, with the nearly two hundred monster movies made during the decade ranging from spectacular to dreadfully tacky.

Spacecraft-oriented monster movies such as *War of the Worlds* (1953) and *Forbidden Planet* (1956) are best known for their special effects. With the launch of the first orbital satellite, Sputnik 1 in 1957, the reality of space travel made the audience skeptical of anything less than realistic.

The classic H. G. Wells invasion story was beautifully updated in the movie *War of the Worlds*. Set in California rather than New Jersey, the movie followed Wells's basic plot with the devastating Martian attack on the cities, the panicked

99. Monsters were big in the movies, and even if they weren't realistic like this one in *The Angry Red Planet* (1960), it didn't matter, because there were plenty more monsters on their way—everything from "It" to the "Thing." Often, monsters in these movies were simply referred to as "Martians," illustrating the way the image of a warlike Mars had permeated society. (Everett Collection)

flight by humans, and the subsequent desire for retaliation by the survivors.

This post–World War II paranoia was perhaps justified. As recent history showed, countries could be invaded without provocation or warning. And secret subversives known as the "fifth column" had undermined Spain's Civil War (1936–39). Many people believed that a fifth column of German agents had invaded England during World War II, and agencies had been established, such as the American CIA, to engage in espionage activities to keep track of what was happening in countries behind the Iron Curtain.

This fear and uncertainty was the fallow ground that produced McCarthyism—and its mendacious efforts to root out the "Red menace"—and the evangelical religious revival led by Billy Graham. The public's fear was nourished by constant propaganda of communists, who were reportedly trying to subvert the God-fearing, free world. And the frightening thing about communists, as everyone knew, was that from the outside they looked just like us.

Movies during the 1950s followed this subtle twist in logic, presenting an alien menace that could masquerade as human. Suddenly our friends, our lovers could be taken over by aliens and become something to be hated and destroyed. Low-budget films seemed to gravitate toward this plot device since it made the special effects for their monsters much easier to deal with. *Invaders from Mars* (1953), *Invasion of the Body Snatchers* (1956), and *I Married a Monster from Outer Space* (1958) were among the most successful of these types of movies. *Invaders from Mars* was one of the earliest and most striking in its use of this technique; in the story the protagonist's parents were the first to succumb to alien control. After all, when it gets to the point that you cannot trust your own mother, who can you trust?

100. This poster for the movie *Invaders from Mars* (1953) shows the Martian as a terrifying, dominating force. People loved monster movies because the movies summed up an instinctual fear of change and satisfied a sneaking suspicion that progress would unleash monsters on our world. (Everett Collection)

Chapter 8

THE MARINER MISSIONS
GO TO MARS

The coming together of two worlds took place at long last on November 18, 1964. Off in the distance was Mars, a bright red glint millions of miles beyond the launch platform at Cape Canaveral on Earth. Scientists looked to Mars as the planet that could hopefully be transformed from an enigma into a friendly territory for mankind.

And at this moment the future that had been imagined throughout centuries of Earth-based observations was very real. Framed against a moon-lit ocean, the Atlas-Agena rocket waited to lift off for Mars with its precious cargo—Mariner 4. This launch was the fruition of countless dreams of crossing the unimaginable distances to our nearest neighbor, the planet most like our own.

The Mariner probe successfully completed its program of taking twenty-two scenic views of Mars, capturing only 1 percent of the surface. Signals raced from Mariner, sending the

101. As seen in this 1960s photograph, the moon is suspended over Cape Canaveral's long row of rocket gantries from which all the Mariner and Viking spacecrafts would be launched. NASA reshaped the Cape, originally a fifteen-thousand-acre sand-spit on the east coast of Florida, into a fascinating, exciting, and at times overwhelming port of blinding searchlights and active launch pads.   (U.S. Air Force)

102. On November 28, 1964, an Atlas-Agena rocket boosted the Mariner 4 probe on its 325-million-mile, seven-and-a-half-month journey to Mars. This probe sent back the first close-up views of a planet other than our own. (NASA)

strange coded dots 134,826,000 miles back to Earth. To cross such distances, the systems required eight hours and thirty-five minutes of flawless operation.

The Mariner 4 photographs were blurry and imprecise compared to the next NASA missions to Mars in the late sixties (Mariners 6 and 7). Yet the images were clear enough to spark a worldwide controversy among astronomers. As the first pictures were retrieved from the shaded dots and converted to photographs, one could almost hear the theories and long-held beliefs come crashing to the laboratory floors like the shattering of so many clay pots. Mars was not what scientists were convinced it would be.

No canals. No artificial waterways. Not a single sign of life. Instead, the photos of Mars revealed an astonishing field of vast craters and countless swarms of lesser craters. Mars looked like the Moon.

Professor Bruce C. Murray of the California Institute of Technology, and a lead member of the NASA team presenting the "Initial Interpretation of Mariner 4 Photography," told the first press conference after the mission that much of the interpretation of Mars was based on only a few photographs.

The Mariner 4 scientists believed that the cratered features indicated that the surface had remained essentially unchanged for billions of years. They determined that Mars was frozen in time and that the surface "must be two to five billion years old, a very ancient surface indeed."

Compared to Earth's weather patterns, with its fierce scrubbing and changing surface, Mars is comparatively sterile. Yet the red planet is far more dynamic than astronomers guessed in the early 1960s based on the Mariner 4 photographs.

There are basic forces at work on the surface of Mars. First, the temperature spans several hundred degrees between maximum and

· ·

103. For 228 days, Mariner 4's ungainly flying windmill of solar arrays had to survive the extreme thermal changes in the space vacuum. Temperatures climbed above 300 degrees and fell to below 200 degrees Fahrenheit during the planetary ship's journey to Mars.   (NASA/JPL)

minimum levels. Soil and rock cannot be alternately baked and frozen for billions of years without changing in density, structure, and makeup. These temperature swings create great amounts of dust. With even minimal movement, dust produces additional weathering, obstruction, and clearing.

There is also the constant rain of cosmic radiation bombardment on the surface of Mars—from the universe and, more particularly, from our own sun. The Sun constantly sends forth boiling electron storms, heavy subatomic particles, and fierce electrical energy fields. These particles whip against Mars at the speed of light. A few billion years of this kind of attack separates chemical constituents from objects, creating new types of material.

And finally, no planet could remain unchanged when it is bombarded by asteroids and meteors as often as Mars.

As scientists at the Smithsonian Astrophysical Observatory subjected the photographs to intense scrutiny and evaluation, they determined that the changing features observed on Mars absolutely could not be due to vegetation. Carl Sagan and James Pollack agreed that the cause of color variations is due to windblown dust.

It is not surprising that the colorful changes seen in the Martian biosphere are not caused by springtime changes. After taking tens of thousands of photographs of Earth, scientists have discovered that it is impossible to detect vegetation growth and decline from even the low-orbit satellites. The fantasy of a lush Martian environment went up in a puff of smoke, and it was agreed that vast, barren deserts cover the entire planet.

WATER ON MARS

Water was the next question, because if there were water on Mars, then there could be life. Astronomers who investigated "Martian weather" through Earth-based telescopes had gathered evidence of a filmy whiteness, which was targeted for close examination.

When Mariner 4 photographed this area, it appeared that frost rimmed the craters. In one of the few successful interpretations of the photographs, scientists were able to predict with

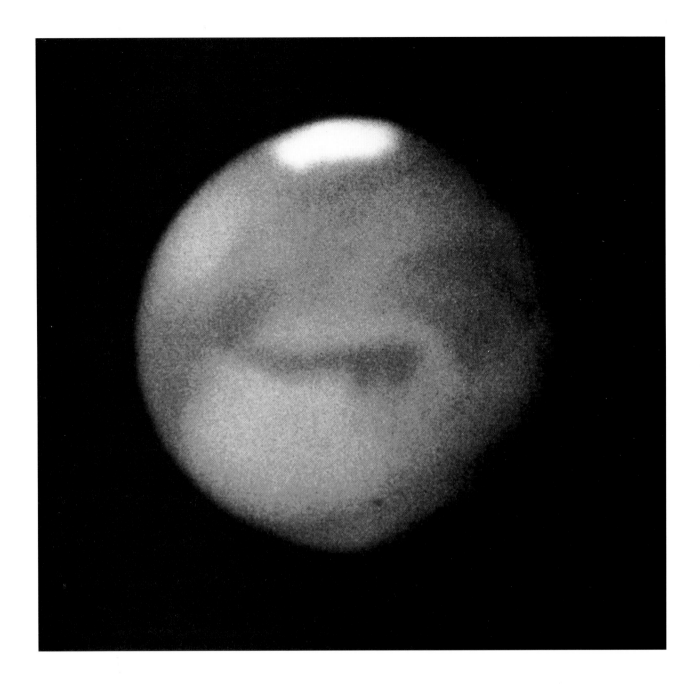

104. As they studied the Mariner photographs, scientists referred back to observations of ground-based telescopes, as seen in this photograph of Mars by Robert Leighton of the California Institute of Technology. Could the large dark areas on the planet's surface have anything to do with vegetation, as once believed, or could they simply be craters hundreds of miles in diameter?

(NASA/Dr. Robert Leighton, California Institute of Technology)

· ·

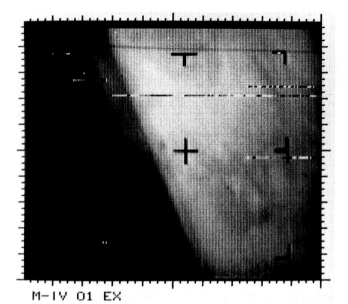

M-1V 01 EX

105. This is picture Number 1 taken by Mariner 4's camera on July 14, 1964—an historic photograph in human exploration. It represents the first time another planet was recorded from a spacecraft. The image was taken across a slant range of 10,500 miles, and, following months of careful study, scientists agreed that one of Mars's few clouds can be seen near the planet's rim. (NASA/JPL)

surprising accuracy that the composition of the frozen gases near the surface consisted of carbon dioxide, among others. And it was theoretically possible for frost to be produced from the rapid temperature changes in even the thinnest atmospheric vapor.

On certain crater walls the frost appeared clearly, brightly reflecting the final rays of the Sun. Yet many major scientists—the foremost being Gerard P. Kuiper from the University of Arizona—insisted that no frost or ice could be indigenous to Mars. He claimed comets had deposited heavy traces of frozen gases that made the crater walls *seem* to be ice coated.

In the midst of such disagreement, it is hardly surprising that the question of canals refused to dissipate—even after studies were made of the photos from Mariner 4. Scientists were hounded at the news conferences by reporters hammering for a direct yes or no about the existence of canals on Mars.

But with only 1 percent of the surface photographed, it was impossible to give the public a concrete answer. NASA officially stated of the twenty-two photographs from Mariner 4, "Their meaning and significance is, of course, a matter of interpretation." One NASA scientific panel on the mission wrote, "Although the flight line crossed several 'canals' sketched from time to time on maps of Mars, no trace of these features was discernible."

In the face of the media circus surrounding the question of canals on Mars, the response of Robert Leighton from the California Institute of Technology perhaps was a bit tongue in cheek. He said that though Mariner 4 had not found any indications of canals, "My canal-gazing friends tell me we picked a very poor place on Mars and a very poor season. So we don't yet know for certain whether there are canals on Mars, but we can say we haven't seen any."

In defense of the existence of canals, Eric Burgess, Fellow of both the Royal Astronomical Society and the British Interplanetary Society, pointed an accusing finger at the notorious Frame Eleven of the Mariner 4 photographs. It shows the area of Atlantis between Mare Sirenum and Mare Cimmerium, spanning 170 miles east to west and 50 miles north to south. One crater is approximately 75 miles in diameter, with multiple lesser craters at least 3 miles in diameter (and presumably even smaller), punched into the surface.

A cursory look at the photograph fails to reveal anything unusual. But on closer examination, like magic, something unexpected appears. Burgess stated that "it clearly indicates a defined linear feature in the position of a Martian canal. In this picture . . . the canal appears as a rift valley thirty miles wide."

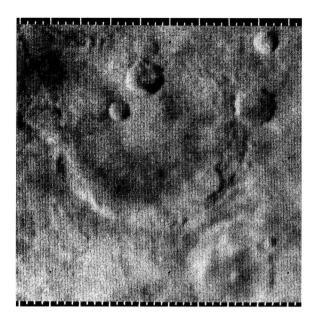

106. The surface of Mars in the Mariner 4 photographs suggested a lunarlike environment. Picture Number 11 covered an area of 170 by 150 miles. Note the single giant crater, the rim walls broken by newer, smaller craters, and above all the linear feature slanting up from the lower left corner—one of the first real indications of the Martian *canali*. (NASA/JPL)

Burgess claimed in his book *Spaceflight* that the observers of the Mariner 4 photographs committed the "unforgivable scientific error" of postulating that Mars had a surface like that of our moon simply because it was cratered. Astronomers did indeed learn a lesson from the Moon that would apply to the Mariner 4 photographs of Mars. Early Earth-based telescopic observations of the Moon showed there were long cracks or gaps in the surface, yet when cameras were sent closer to the surface, instead of the sharp cuts, slices, grooves, and other cracks, we found ourselves staring at lines of craters. What appeared from a distance to be a solid unbroken line became separate depressions.

Schiaparelli, Lowell, Arnold, Fox, and all the other astronomers who mapped the canals of Mars committed to paper precisely what they saw: lines that could have been either canals or vegetation bordering the irrigation waterways, visible for brief seconds, but seen again and again by different observers. None of the early astronomers were wrong except in stating they saw *canals* rather than crevices and craters, the *canali* that Schiaparelli insisted could be natural topographical features.

A year before Mariner 4, E. C. Slipher of the Lowell Observatory had warned would-be Mars observers that their latitude of observation was directly related to what they may see. As a final note he added that years of experience observing Mars

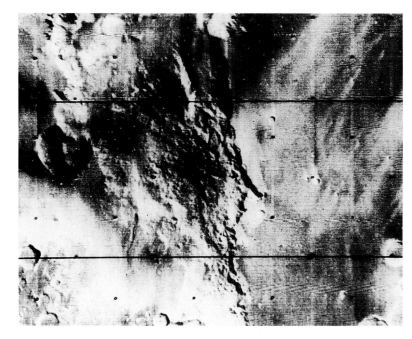

107. This Mariner 6 photograph shows Martian features that would later become familiar to scientists. However, their first look at the distinctive, chaotic terrain of Mars amazed the Jet Propulsion Laboratory's Mission Control in California. (NASA/JPL)

showed that "every skilled observer who travels to the best available site has no great difficulty in seeing and convincing himself of the reality of the canals."

With Mars, perception had always counted for far more than the eye. We often believe what we wish to believe and it takes only a slight push for us to go over the edge. You can prove this to yourself. Draw several straight and curving lines lightly in pencil on a sheet of white paper. Along the thin pencil lines, ink in heavy black dots. Keep the dots close together but not touching. Have someone hold up the paper and walk away slowly. If you watch the paper the entire time, it will not take long before the separate dots vanish to be replaced with solid lines.

Welcome to Mars.

CULTURAL REVOLUTION

The decade of the sixties not only ushered in the era of spaceflight; it was infamous as a time of radicalism and social change. As the mass of baby boomers reached their late teens, their influence on art, sex, politics, and human interactions sparked a division in Western society, especially in the United States, where perceptions of life and reality were rocked.

The mythology of Mars changed once again with the realization that the planet would remain a mystery despite the technological ability to send powerful spacecraft zooming as close as twenty thousand kilometers. This frustrating combination of knowledge and ignorance once again sparked an ironic view of Mars reminiscent of Swift's Gulliver novels during the Enlightenment and Lucian's tales of Mars during the decline of the Roman Empire.

The event that had the most impact on the cultural revolution was the war in Vietnam. The schizophrenic view of the sixties was exemplified by politicians' refusal to acknowledge that the war was a lost cause. Instead, it was called a "police action," and the body count was reported on the news every evening.

The assassination of John F. Kennedy in 1963 had also contributed to general feelings of paranoia. Authors responded with dark stories such as Stanley Kubrick's film *Dr. Strangelove, or How I Learned to Stop Worrying and Love the Bomb* (1964) and Allen Drury's novel *The Throne of Saturn* (1971).

Drury's description of Russian scientists who attempt to sabotage America's first manned expedition to Mars was ironically told, yet true to real life. Even writers who weren't interested in the genre aspects of sci-fi, such as David G. Compton, used Mars to examine the effects of contemporary society. In *Farewell, Earth's Bliss* (1966), Compton examined the struggles endured by social misfits who are transported to Mars.

108. "I was a victim of a series of accidents," the hero proclaims in *Sirens of Titan* (1959). "As are we all." Kurt Vonnegut took the ultimate leap in absurdist science fiction in this novel, with the protagonist encountering standard genre devices such as time travel, spaceflight, and an invasion from Mars. Yet the lyrical, deadpan delivery of the wealthy protagonist being brainwashed into innocence on Mars goes beyond the traditional sci-fi realm.

(From *Sirens of Titan* [jacket cover] by Kurt Vonnegut. Used by permission of Delacorte Press/Seymour Lawrence, a division of Bantam Doubleday Dell Publishing Group, Inc.)

109. Mars first invaded television with Marvin the Martian, zipping around the galaxy with a green dog named K-9 in the spaceship *Martian Maggot*. Warner Brothers first introduced Marvin to the public in a Bugs Bunny episode, "Haredevil Hare," in 1948. In this animation cell, Marvin is arrogantly letting Daffy Duck know that Mars has claimed this piece of real estate.

(Everett Collection)

POP CULTURE

Sci-fi has always been a "popular" medium, and with the growing power of young people in the sixties, writers and producers had to take into consideration the fact that most of their audience was far more interested in the miniskirt than in serious cultural debates.

Comic takes on Mars had begun as early as 1955 in Fredric Brown's *Martians Go Home!* Brown's Martians were little green men who must be imagined out of existence by the writer who created them. In an ironic twist, the writer himself discovered that he was a figment of someone else's imagination.

Sci-fi imagery had already entered rock music in the fifties with songs like "Flying Saucers Rock and Roll," "Martian Hop," and "Purple People-Eater." But the psychedelic music of the sixties did even more to popularize science fiction. With San

Francisco rock groups playing to the drug revolution, the Grateful Dead put out albums called *Aoxomoxoa* (1969) and *From the Mars Hotel* (1974) as well as performing eerie improvisational pieces like "Dark Star."

David Bowie also helped make sci-fi fashionable with his character, Ziggy Stardust, and the band called Spiders from Mars, on their album *The Rise and Fall of Ziggy Stardust* (1972). His rock song "Space Oddity" (1969) told of astronaut Major Tom who was so taken with space that he decided to wrench himself away from family and Earth forever. The music starkly conveyed people's passion for ultimate freedom, and, fittingly, it was used during the televised coverage of the first Moon landing.

One movie, *Robinson Crusoe on Mars* (1964), was a classic sixties combination of two contrasting trends—serious and silly. The talented director of this movie was Byron Haskin, who had created several competent sci-fi films the decade before, including *War of the Worlds* (1953). For the 1964 movie, Haskin's vision of a futuristic Mars was based on Rex Gordon's novel *No Man Friday* (1956), which in turn was a futuristic version of Defoe's classic novel *Robinson Crusoe* (1719). Though the science was sometimes shaky in Haskin's film, the script resembled Gordon's story in its pessimistic tone of survival that was typical of the fifties.

110. Hollywood responded to the demand for satiric takes on Mars by putting out hundreds of silly, colorful sci-fi movies and television series. *My Favorite Martian* (starring Ray Walston) mimicked the success of the Batman series in its spoof on the sci-fi themes that had become part of pop culture. (Everett Collection)

111. In *Robinson Crusoe on Mars* (1964), the astronaut is the lone survivor when his spaceship crashes on Mars. With only a monkey for company, Crusoe's sanity begins to slip—until the timely, melodramatic arrival of an alien Friday. The desolate environment was hauntingly filmed in California's Death Valley, the perfect setting. (Everett Collection)

Philip K. Dick was the writer who best captured the unique juxtaposition of reality and perception that ruled during the sixties' cultural revolution. His depictions of consensual reality usually reflected the empty irony of modern culture, particularly in how we destroy that which is most valuable to us. Dick's work was the forerunner of the most recent genre shift into cyberpunk.

As in *Martian Time-Slip* (1962), Dick's stories were both funny and terrifying. In this early version of Mars, Dick envisioned the plant as contaminated by schizophrenic perceptions, trapping the weak characters in their own narrow lives. He explored the uses of hallucinogenic drugs in *The Three Stigmata of Palmer Eldritch* (1965), thereby making a desolate Mars bearable for a small group of colonists.

The best of Dick's thought-provoking books was *Do Androids Dream of Electric Sheep?* (1968; made into the movie *Blade Runner*, 1982). Mechanical androids were used to exploit Mars, yet when they escaped to Earth they were hunted down by a futuristic bounty hunter. The most poignant aspect of the novel was the empathy human beings felt for animals—most had died out and were replaced by mechanical creatures in order to satisfy the human need for contact. In the ultimate irony, the protagonist discovered that the new messiah—on whom all the hopes of humanity had been staked—was a fake.

Creating messiahs was a favorite ploy in sci-fi fiction of the sixties. In Roger Zelazny's caustic, wisecracking story "A Rose for Ecclesiastes" (1963), a human messiah ultimately redeemed the Martians despite his Loki/Trickster characteristics.

Even better known is Robert Heinlein's *Stranger in a Strange Land* (1961),

112. Most of Philip K. Dick's novels were bleak as far as the human condition is concerned. The essence of *Do Androids Dream of Electric Sheep?* (1968) was captured by Ridley Scott in *Blade Runner* (1982) even if the exact story line was not followed. In Dick's original, the protagonist's longing for a real animal, not a mechanical construct, permeates the tale with bittersweet sadness.

(From *Do Androids Dream of Electric Sheep?* [jacket cover] by Philip K. Dick. Used by permission of Doubleday, a division of Bantam Doubleday Dell Publishing Group, Inc.)

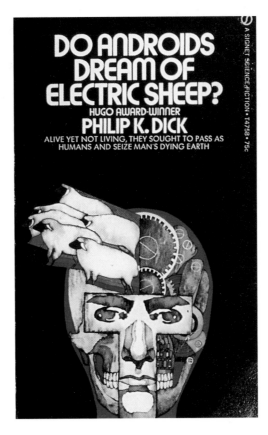

DO ANDROIDS DREAM OF ELECTRIC SHEEP?
HUGO AWARD-WINNER
PHILIP K. DICK
ALIVE YET NOT LIVING, THEY SOUGHT TO PASS AS HUMANS AND SEIZE MAN'S DYING EARTH

A SIGNET SCIENCE-FICTION • T4758 • 75c

THE MARINER MISSIONS GO TO MARS

113. This is the spacecraft used for the Mariner 6 and 7 missions. It is much larger and heavier than Mariner 4. Equipped with two television cameras and a large disk-shaped antenna, the new Martian ships could transmit their information back to Earth with a speed of 16,200 bits per second instead of the slow 8.3 bits per second as the hampered Mariner 4. (NASA/JPL)

the first of his stories that centered around sexual emancipation. The protagonist, Valentine Michael Smith, who was a human raised by Martians, brought a religious philosophy of ultimate acceptance back to Earth.

Heinlein is often considered to be the most important writer of genre sci-fi. He was prolific, writing for juveniles as well as adults, including *Red Planet: A Colonial Boy on Mars* (1949) and *Podkayne of Mars: Her Life and Times* (1963). His writing style was a confident, easy blend of contemporary slang and techno-speak. The story narrative of the characters took precedence over the sci-fi elements, with the innocent Smith turning out to be a frighteningly powerful messiah figure who discovered the secret of transcending death.

Philip K. Dick, Roger Zelazney, and Ursula Le Guin were among the New Wave of literary sci-fi writers who raised the level of exploration of psychology, sociology, and linguistics in the genre. These writers, along with genre greats Isaac Asimov and Robert Heinlein made the most of the freedoms of the sixties to bring their sci-fi imagery to new life.

The world was gripped by a reinvigorated Mars fever after the Mariner 4 mission. As the 1960s came to a close, the United States prepared to return to Mars armed to the teeth with improved scientific equipment and a more powerful rocket booster — the Atlas-Centaurs of Mariners 6 and 7.

Far more ambitious than Mariner 4, the new vessels were intended to photograph at least 20 percent of the planet as they flew past. They were large and heavy spacecraft with even more scientific sorcery imbedded in their systems than the now obsolete Mariner 4. Each one carried scan platforms to make maximum use of sophisticated equipment: two television cameras, an infrared spectrometer, an ultraviolet spectrometer, and an S-band (radio) occultation experiment.

On February 24, 1969, Mariner 6's rocket thundered from launch pad 36B at Cape Canaveral. Fourteen minutes later, separated from its boosters, Mariner 6 ripped away from Earth at

114. During its accelerating approach, Mariner 7 shot this full view of Mars. The planet's polar ice cap is clearly visible despite the shadow shrouding Mars's southern hemisphere. The cloud formations and haze are seen over a volcanic and cratered surface. Mariners 6 and 7 shot more than two hundred photographs as they approached the red planet.　　　　　　　(NASA/JPL)

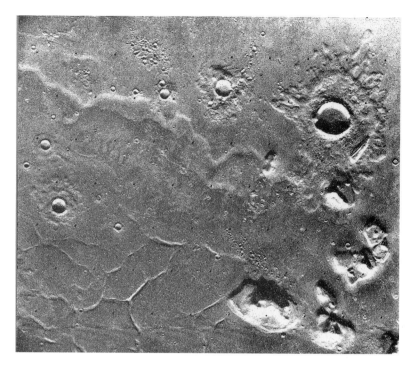

115. This photograph taken from Martian orbit shows some of the smallest craters visible to the Mariner spacecraft. But even the smallest crater visible in this picture is approximately three miles in diameter. The most famous meteor crater on Earth is Sunset Crater in northern Arizona, 4,000 feet from rim to rim, less than a square mile in total area. (NASA/JPL)

twenty-five thousand miles per hour. Then on March 27, 1969, the second half of the dual mission was launched as Mariner *7* roared upward. Though a month separated their launches, by the time they reached the vicinity of Mars the twin spacecraft were separated by a flight time of only five days.

The Mariners could each store thirty-five times more data than the Mariner 4 mission five years earlier. Mariner 6 snapped fifty photographs during its approach and twenty-five more photos during the surface-skimming pass. A handful of days later, Mariner 7 raced by Mars, adding 124 pictures during its approach and minimum distance flyby.

CHAOTIC TERRAIN

It was typical of Mars that a somewhat confusing image of the planet was created by the Mariner missions. As fast as one new element of the surface or weather conditions was found and seemingly confirmed, something new would arise to confound earlier findings, establishing new mysteries for scientists on Earth.

Mariner 4 had promised a world of craters with its paltry 1 percent of photographic coverage. Scientists had no doubt that

116. Mars may have craters like Earth's Moon, but they are not the same. Mariners 6 and 7 revealed that some sort of weathering process was steadily changing the surface. Features, such as the large crater seen in the right of this photograph, were clearly being eroded. (NASA/JPL)

they would find craters on Mars, perhaps hundreds of thousands of them. Most scientists also had no doubt that there would be no evidence of canals found on Mars.

The initial photographs contradicted those taken by Mariner 4 with a number of "major oases" located in the exact positions where decades of observers had charted "canals." Though these dark areas were not oases in terrestrial terms of water and vegetation, they were found to be single, dark-floored craters such as Juventae Fons, or groups of craters, Oxia Palus.

The twin Mariners confirmed craters up to three hundred miles in diameter (Nix Olympica), yet craters were not found with any uniformity across the world. Not all photos produced the clarity desired because much of the terrain of Mars is chaotic—as NASA describes it, "irregular, jumbled topography, reminiscent of the slumped aspect of a terrestrial landslide."

There are also enormous areas of featureless terrain. In the high desert floor of Hellas, with an especially bright reflection in the cameras, the floor of the crater proved unexpectedly smooth. Picture resolutions down to a thousand feet showed a surface area as flat as dry lake beds on Earth—which are so flat and smooth they are used as airfields for high-performance

117. The clarity of Mariner 7's photographs created as much confusion as ever. Gerard P. Kuiper, world-renowned scientist from the University of Arizona, refused to believe that Mars's thin atmosphere could produce frost. Dr. Kuiper argued that a comet with millions of tons of frozen water and gases must have splattered like a hydrogen bomb into the Martian surface, leaving this crater and its frozen rim visible in the upper left corner of the photograph.   (NASA/JPL)

aircraft and rockets, without any need for surface improvement or paving.

These smooth areas bewildered scientists, particularly the area of Hellas, which covered twelve hundred miles. Anything a thousand feet in diameter or larger stuck out like a sore thumb on the photographs. Yet nothing was seen. The scientists admitted, "There is no way in which Hellas could be sheltered from impacts by meteorites." The preliminary conclusion was that the material on the floor of Hellas "responds more quickly than any other Martian materials to whatever processes of erosion exist on Mars."

With new evidence that Mars was not "dead" as predicted, based on Mariner 4 photographs, scientists focused on the polar caps. Craters studded the boundary of the southern cap, and several seemed to be partially covered with "snow." Tentatively, the NASA Jet Propulsion Laboratory (JPL) team announced that the polar regions were covered by drifts of some sort of "material," which could be several feet in thickness.

The early indications of atmospheric haze found in the photographs from Mariner 4 were confused by these images. Some scientists estimated that the haze layer extended between ten and thirty miles above the surface, while others insisted it clung close to the surface. Blue, red, and green filters indicated that the atmosphere contained small particles of aerosol, though some scientists disagreed, saying the haze most likely consisted of solid particles of carbon dioxide the size of dust motes.

Infrared radiometer scanning reported a severe drop in temperature over the polar caps, and after a full sweep using different instruments, it was decided that the polar caps consisted of carbon dioxide. The presence of carbon dioxide was fully confirmed, being found in every area covered by the Mariner instruments, while large traces of carbon monoxide were also found.

Some of the photographs did reveal "several diffuse, bright patches suggestive of clouds near the polar cap edge . . ." but none of them were conclusive. Yet in the south polar region scientists were frustrated by areas that seemed to be covered with

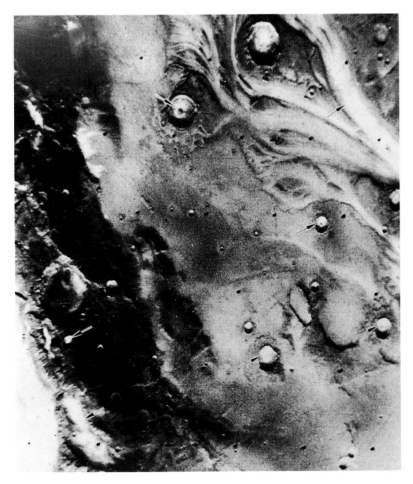

118. As Mariner 7 swept across the south polar region, scientists refused to believe they were seeing "fog" filling the craters below. Mariner 4 photographs had convinced them fog was not possible on Mars. But the evidence was captured on film, to be confirmed by later NASA probes. (NASA/JPL)

fog, while others claimed it was simply fogging of the film. The general consensus of astronomers post–Mariner 4 was that fog could not exist on Mars. But the more the Mariner 6 and 7 photographs were studied, the more convinced scientists were that this was pure, unadulterated Martian fog.

Fog meant there could be free water in the atmosphere. Surface features exist that could indicate water once flowed across the surface of Mars. But the scientists saw nothing that was conclusive regarding ancient waterways. They would have to see more, and in order to do that they would need more time. The next Mariners to be launched from Cape Canaveral would orbit Mars rather than fly past.

Chapter 9

THE REALITY OF MARS IN THE 1970s

The eagerly anticipated launch of Mariner 8 on May 8, 1971, proved to be a sharp disappointment. Less than twenty seconds after a flawless liftoff, a small diode in the guidance system failed. Balance and directional control vanished, and a beautiful spacecraft became a mindless, tumbling object, falling into the Atlantic Ocean.

What made it even worse was that it failed in front of the entire world—especially since international competition had become the driving force behind the U.S. space program. Since 1962, the Soviet Union had also set its sights on the red planet. Their first shot to Mars took place before Mariner 4, but the aiming system for the radio antenna malfunctioned, and the Russian probe went dead.

Following this, the Russians fired another *six* spacecraft, and again chalked up *six* failures. Mars seemed destined to be unreachable for the Russians. Their consolation lay in repeat

119. As seen in this NASA photo, the darkness covering launch pad 36A at Cape Canaveral was pushed away by the brightness of ignition on May 8, 1971. The Atlas-Centaur rocket lifted Mariner 8 away from Earth only to have its Centaur stage fail, sending the Mars-bound spacecraft tumbling down into the ocean.   (NASA)

. . . .

missions to Venus, which were spectacularly successful, including parachute probes through the dense atmosphere and actual landings on that hellish world.

Then on May 19, eleven days after the failure of the U.S. Mariner 8, the Russians sent a heavy Mars 2 spacecraft to Mars. More than five tons in weight, the Russians identified their shot as an "automatic interplanetary station" beginning its 290-million-mile flight to Mars. Eight days later, the Russians launched another enormous spacecraft, Mars 3.

Three days after that, on May 30, 1971, Cape Canaveral launched Mariner 9. It was scheduled to reach Mars and enter orbit (not simply fly by like the earlier Mariners) slightly ahead of the Russian crafts.

American and Russian scientists established a hot line between the two countries to assist one another in getting the best results from the probes. Using the orbital numbers gathered by Mariner 9, the Russians were able to tweak their own spacecraft into perfect position to enter orbit.

Yet age-old suspicions about the "Reds" revived in the middle of the Mariner 9 mission, severely cooling relations between international scientists. Just before Mars 2 was to fire its engine for orbital insertion, the craft ejected a small capsule carrying metal banners and emblems to celebrate "the first hard landing" on Mars. For two days the Russians refused to answer whether the capsule had hit Mars or missed completely—all of which left American scientists, as one said, "highly ticked off and ready to tell the Soviets where they could stick their whole program."

The Russian government was more open about the Mars 3 mission to land an instrumented probe on the surface, reporting that the probe "entered into the planet's atmosphere,

120. The media reported a severe strain between American and Russian scientists as to who might reach Mars first, and who would obtain the best or most data. This promotional poster for *Capricorn One* (1977) captures the spirit of competition that spurred on space exploration during the seventies. (Everett Collection)

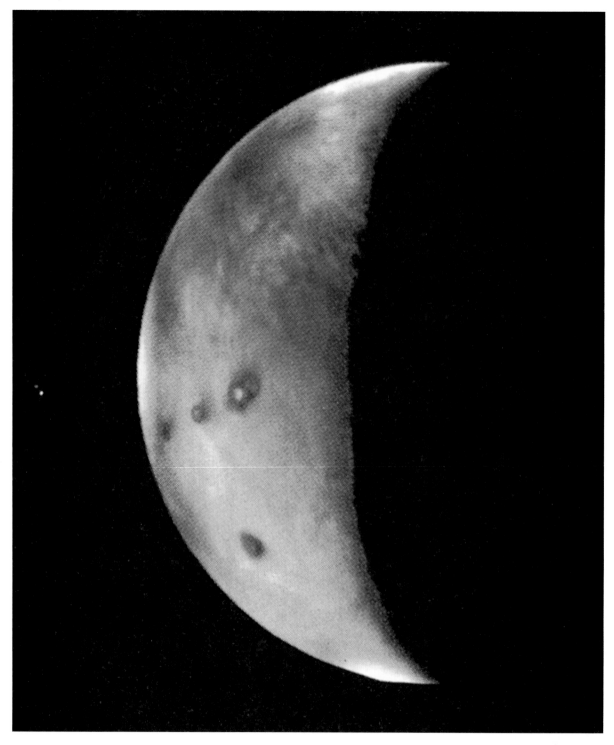

121. One of the first photographs taken by Mariner 9 revealed the largest volcano ever seen on a planet. Larger than the state of Missouri, Olympus Mons is only one of four volcanoes that form a chain across the surface, including Arsia Mons, Pavonis Mons, and Ascraeus Mons. **(NASA/JPL)**

parachuted down and softly landed in the southern hemisphere of Mars between the Electris and Phaetonis regions."

NASA monitored the radio transmissions that Mars 3 sent back to Earth, hoping that the lander would produce close-up video or photographs of the surface. But the Russians, in an uncharacteristic move, admitted that they had suffered a drastic failure with their equipment. The lander had only sent "some" video signals, which were "brief and suddenly discontinued." In short, another technological failure struck the Mars missions of the early seventies, and a golden opportunity was lost. It would not be until the Viking launch in 1976 that Earth would finally get a good look at the surface of Mars.

GOVERNMENT PARANOIA

Though the public was no longer quite so hysterical over the Cold War with the Russians, by the seventies the general belief that a nuclear attack was possible at any moment had became an integral part of American culture. Schools were long past holding drills to send children under desks in the event of a bomb — but that was because the obliterating reality of a nuclear holocaust was known all too well.

The post–cultural revolution era marked a shift in the focus of public paranoia; what began as an externalized fear of

122. As astronauts, James Brolin, O. J. Simpson, and Sam Waterston are fully prepared to blast off for Mars in *Capricorn One* (1977), but are later terrified to learn that they have been duped into a faked mission, and in order to save their lives they must act out the exploration of Mars in front of the television cameras.   (Everett Collection)

invasion and communism turned inward, and the American people began to distrust their *own* government. First there was the dramatic betrayal of public confidence with the 1972 break-in at Watergate, which climaxed with President Nixon's resignation. Then there was the dismal withdrawal from the "police action" in Vietnam in 1975—with officials continuing to deny it had been a war, and a failure, to boot.

American culture was faced with the unacceptable fact that the "American way" had not succeeded overseas. Politicians responded by trying to shove the whole nasty business out of sight and out of mind, while science fiction blamed the bureaucratic system for all of society's ills. The majority of the sci-fi films in the 1970s were particularly gloomy, reflecting the general feeling that individuals were at the mercy of unseen forces.

Conspiracy theories were so common that NASA actually cooperated in the production of *Capricorn One* (1977), a film about government paranoia. In this movie classic, a U.S. manned mission to Mars was called off due to a faulty life-support system, and NASA faked the mission by launching an unmanned probe. In order to head off a media disaster and subsequent cut in funding, the astronauts were blackmailed into appearing on a movie set where they pretended to land on Mars. NASA then tried to kill the astronauts in order to keep the whole thing a secret, and the subsequent chase and escape was more typical of a standard thriller than science fiction.

123. This is the windmill-shaped Mariner 9. Unlike the Atlas-Centaur that failed Mariner 8, Mariner 9's rocket worked flawlessly and the 1,245-pound planetary ship was placed on the precise course with the perfect speed needed for its 247-million-mile journey to Mars.
(NASA/JPL)

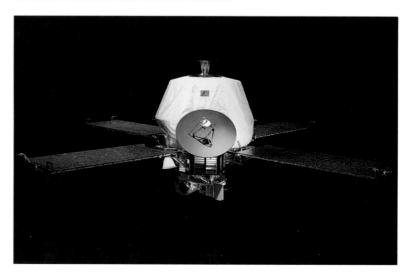

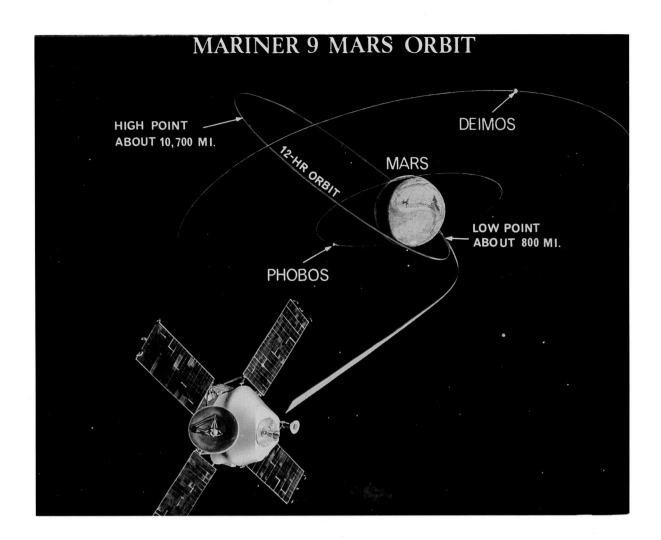

MARINER 9 MARS ORBIT

HIGH POINT
ABOUT 10,700 MI.

DEIMOS

12-HR ORBIT

MARS

LOW POINT
ABOUT 800 MI.

PHOBOS

MARINER 9

There were bigger problems than Cold War competition waiting for the scientists who were studying Mars with Mariner 9. When the probes (including Mars 2 and Mars 3) entered orbit, taking pictures and scientific measurements, they immediately confirmed the existence of weather on the planet. The entire surface was buried beneath a huge dust storm, and all the photographic images were of roiling dust.

This time, however, it was not just a swift flyby. Mariner 9 was the first American spacecraft to go into orbit around a planet other than Earth, and it saved the mission in more ways than one. All three of the spacecraft were there to stay and they waited until the dust storm began to dissipate.

124. As seen in this artist's illustration of its mission, Mariner 9 completed its precise journey to Mars and slipped into orbit on November 13, 1971. Its closet approach to Mars was 868 miles and the high point of its orbit was 10,655 miles. (NASA/JPL)

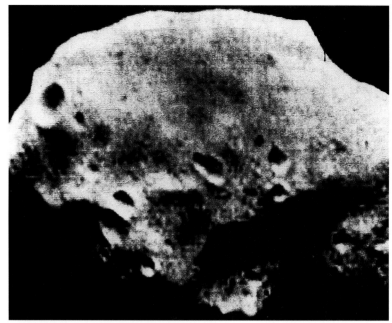

125. Phobos, the small Martian moon, was photographed during Mariner 9's thirty-fourth orbit. Irregular and battered, Phobos is barely fourteen miles in diameter, yet it is billions of years old. Scientists have determined that the moon must have an unusually great cohesive strength to survive in the Martian orbit. (NASA/JPL)

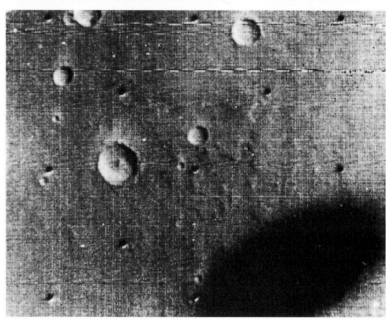

126. The shadow of Phobos appears on the lower right corner of this photograph, cast by the slanting rays of the Sun. The photos taken by Mariner 9 were the first ever of the surfaces of the two Martian moons. (NASA/JPL)

130. In the fractured volcanic tablelands of Noctis Lacus we find one of the most sensational features of Mars. Laid out as if on a flat table is an intricate network of magnificent canyons covering an area 336 by 264 miles. This surface sculpture looks for all the world like a giant candelabra that could burst into dazzling light at any moment. (NASA/JPL)

· ·

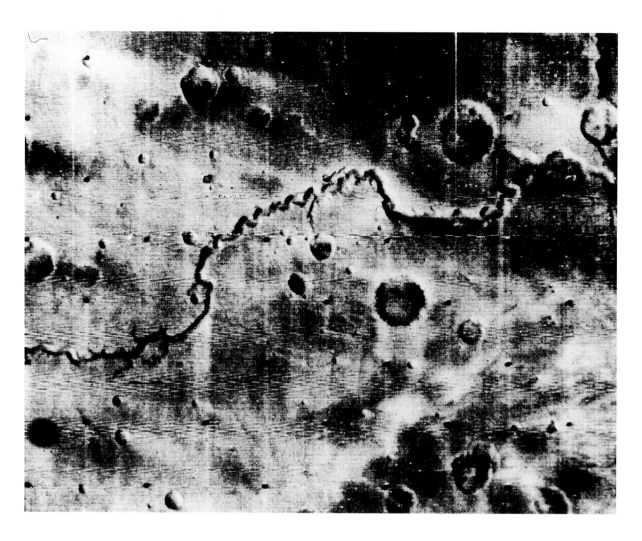

131. Mariner 9's photograph of this "arroyo" snaking across the sinuous valley was just one of the spacecraft's many images that made the Martian surface a thousand times more interesting than the limited detailed photographs sent back by the early Mariners.   (NASA/JPL)

WEATHER ON MARS

Scientists examined the terrain captured in the aerial photographs for any evidence of the existence of water on the surface. Previous Mariner missions had already proved how tricky evidence based on photographs could be, but everyone seemed willing to speculate—as long as their statements were not taken as fact.

The one sure thing on Mars was the proof of enormous, ancient volcanoes. Scientists agreed that in the distant past Mars had once been extremely active, wracked by searing internal fires and fierce volcanic eruptions, with all the attendant release of gases and dust into the atmosphere. Somewhere in there, many scientists reasoned, there must have been water.

The polar caps received intense attention from all three

orbital probes. It was proved that the caps consisted of mainly carbon dioxide, yet some scientists held out the remote possibility that they also held traces of water. For decades observers on Earth had recorded the seasonal cycles of the polar caps. As the summer heat intensifies, the northern polar cap shrinks and fills vast areas with white vapor. This veil can become brilliant with the reflected sunlight. Many of the Mariner 9 images recorded the existence of Martian clouds. Yet after the contradictory findings of Mars, the official NASA report cautiously stated there were "several diffuse, bright patches suggestive of clouds near the polar-cap edge."

Scientists who supported the theory of clouds on Mars pointed to the powerful forces that generated enough energy to cover the entire planet with dust. This sort of weather was taken as confirmation of clouds, whether they were banks of dust,

132. This color detail of the north polar cap of Mars reveals the dark, undulating line of a giant cliff about 1,500 feet high. Regular layers averaging 150 feet in thickness are highlighted by occasional white patches of frost—a combination on Earth that indicates erosion by ice ages caused by the periodic changes in planetary orbit. This cliff is apparently an erosional feature, with dunelike areas (the rippled texture) possibly formed from material sloughed off by the layered terrain. (NASA/JPL)

water vapor mixed with heavier concentrations of carbon dioxide, or yet some other unidentified materials.

Temperature readings were also taken, but these figures were often misleading. Journalists depicting Mars read that the equatorial summer temperature reached a balmy 70 degrees Fahrenheit. Yet these readings were taken in the soil, recording the amount of solar radiation striking and being absorbed by the surface. It had nothing to do with the ambient air temperature. Without a significant atmosphere to retain the heat, the temperature would always be below freezing.

Solar radiation also is a significant hazard on the surface of Mars because of its thin atmosphere. Mariner 9's findings led to the conclusion that conditions on Mars are much more like those on the terrestrial Moon than on Earth.

An unprotected person on the surface of Mars would receive a radiation dose as high as one hundred times our daily exposure on Earth. It would be like standing downwind from a nuclear testing ground in the American Southwest and being subjected to damaging fallout. Pressure suits and personal shielding could provide adequate protection for weeks, but for longer periods of time the outlook would be fairly grim.

Yet the abundance of information sent back by Mariner 9 prompted scientists to have a higher confidence in Mars as a future home for a human population. They were quick to point out that Mars *as it is now* would certainly pose radiation dangers. But once proper shielding was in place, including a dense atmosphere, the inherent danger would be diminished.

At a press conference,

· · · · · · · · · · · · · · ·

133. This photo startled scientists who had no idea such features existed on Mars. The Phoenicis Lacus plateau lies nearly four miles above the average elevation of Mars and its "elephant hide" surface is made up of broken and fractured volcanic deposits. The individual fault valleys (the wrinkles) are about 8,000 feet across. (NASA/JPL)

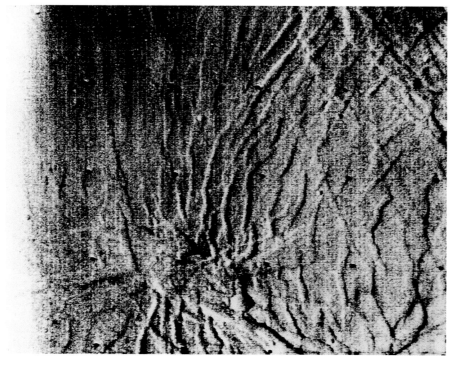

134. One of the most fascinating areas on Mars are the plateaus between Ophir Chasma and Candor Chasma. As seen in this startling color photograph, the enormous, streamlined plateaus bridge the gap between the two canyons, possibly indicating wind erosion on a very large scale. It has also been suggested that a lake (or sea) once existed in the north section of Ophir Chasma until the canyon wall was breached southward to unleash an enormous flood. (NASA/JPL)

one JPL scientist laughed at the suggestion of the extreme difficulties in terraforming Mars to meet the basic necessities of life. "I'll guarantee you," he said. "Just getting to Mars is a great deal tougher than shifting things around on that world to meet the physiological needs of the first settlers."

As usual, sci-fi writers took up where the scientists left off, accepting the challenge of making Mars a viable home for man. Whereas fantasy-based sci-fi creatures had no need to prove their environmental settings, realistic writers had to accept the fact that Mars was deadly to humans.

James Blish originally came up with the term "terraforming" in a group of stories that became *The Seedling Stars* (1957). As Blish saw it, either we would have to change the planet to make it habitable—that is, terraform it—or we would have to change to suit the environment—pantropy.

One of the earliest examples of terraforming is found in a 1909 fantasy novel by James B. Alexander, *The Lunarian Professor and His Remarkable Revelations Concerning the Earth, the Moon, and Mars; Together with an Account of the Cruise of the Sally Ann.* Though the protagonist woke to find his futuristic journey by antigravity was all a dream, his encounters with the terraforming process of Mars were intriguing and plausible.

Frederick Turner's first sci-fi book (he was already a respected poet) was also set on a terraformed Mars in the distant future. *A Double Shadow* (1978) straddled the conflicts between the two strands of human evolution—one on a thriving Mars and the other on a dying Earth. Turner's later work, *Genesis: An Epic Poem* (1988), was even more successful in imagining how Mars could one day become habitable for humans.

Yet the most intriguing novel of the decade involved human adaptation to Mars. Frederik Pohl's *Man Plus* (1976) followed Blish's general notion that it would be easier to make colonists fit the planet rather than altering an entire world.

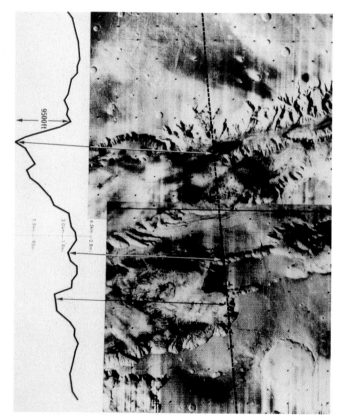

135. A down-faulted valley in Mars's Vallis Marineris canyon seemed to dredge up the ghosts of enormous rivers rushing across the Martian surface some fifty miles wide. It was so unique to this planet that scientists were baffled. Paralleling the rim of the valley is a fluted central ridge. If we could remove the water from the Atlantic Ocean we would find the same type of feature, known as the Atlantic Ridge. (NASA/JPL)

Whereas Blish focused on bioengineering in his seminal novel, *The Seedling Stars*, Pohl came up with a far more ruthless method by turning man into a machine.

Like most of the works of the era, *Man Plus* was ironically grim. It was considered to be a critical reexamination of the theme Ray Bradbury presented in *The Martian Chronicles*—namely, that the Martians were human, but with the new scientific data from the Mariner missions readily at hand, the public was now aware of the hostility of the planet's environment. Pohl's protagonist was believably transformed into a cyborg, in the end fit to live only on Mars, having established a future there for mankind.

136. The diffused markings in this Mariner photograph offered many headaches for scientists. It was images such as these, indicating fluid erosion reminiscent of Earth's terrain, that prompted a fictional response that man could terraform Mars. If water once existed in such abundance, writers reasoned, then it could exist once more. (NASA/JPL)

Chapter 10

LANDING ON MARS: THE VIKING MISSION

After Mariner 9, it was clear that an advanced civilization capable of building Schiaparelli's canals had never existed on Mars. Even sci-fi writers who tried to include ancient Martian cities in their stories were not successful. The same year of the launch of the Viking probes, Peter Edwards's *Terminus* was muddled by the discovery of an ancient city on Mars, which distracted from the development of the political situation in his post-holocaust world.

Even if there were no ruins of monumental structures on Mars, most people believed there must be *some* sort of life on the planet. Though scientists had ruled out the existence of larger organisms, many suspected that in those vast sands and deep crevices there might be something that wriggled or twitched. And at this point, a blade of alien grass or a lumbering Martian insect would have gone a long way to proving the possibility that advanced life was developing on a distant planet in our

137. Two identical Viking planetary ships were launched in 1975. From the tens of thousands of photographs to the chemical analysis of air and soil, this successful mission provided virtually all the scientific data we know about Mars today. (NASA)

galaxy, if not in our own solar system. As always, the search for life led to the search for others that were like ourselves.

Photographs were not enough anymore. To find indications of life, a probe would have to land on Mars. So the last great probe missions of the Golden Age of NASA were launched, with Viking 1 departing Earth on August 20, 1975, and Viking 2 following on September 9, 1975.

To complete the dual mission of orbital surveillance and ground research, seventy-two of the world's best scientific minds formed into thirteen teams. These teams preplanned dozens of complex scientific investigations that were carried out by thirteen separate sophisticated instruments on board both the landers and the orbiters.

The Vikings far exceed their planned mission life: Viking 2's lander ceased transmissions in 1980, while Viking 1's lander lasted until 1982.

· ·

138. Long before the first of two Viking landing craft touched down on the Martian landscape, the spacecraft had to be tested and retested in a similar environment on Earth. Viking project scientist Gerald Soffen is seen here testing a Viking lander prototype in the California desert.

(Gerald Soffen)

139. NASA had extraordinary teams working together in the most comprehensive scientific exploration thus far of the planet Mars. The focal point for all flight operations was in this Jet Propulsion Laboratory Mission Control Center operated by the world-famous California Institute of Technology. (NASA/JPL)

140. This is the Viking spacecraft as it looked when it went into orbit around Mars. The capsule underneath the Viking orbiter contained the spectacular Viking lander. The large disk antenna was used to transmit and receive data between Mars and Earth. (NASA/JPL)

VIKING ORBITERS

Two Viking orbiters went into orbit around the planet in 1976, working perfectly as photo platforms and receiving stations, relaying transmissions from the landers back to Earth. These orbital platforms vastly increased the planetary-wide views,

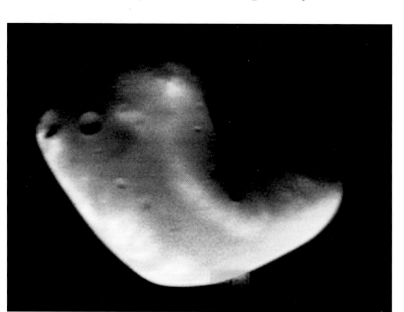

141. The Candor Chasma, part of Vallis Marineris, leaps to new clarity through the Viking photographs. The entire "grand canyon" of Mars stretches 3,100 miles with a maximum width of 62 miles. Earth's Grand Canyon, by comparison, is only 217 miles long and 11 miles wide. (NASA/JPL)

142. This color photo of Deimos shows the surface to a resolution of a few hundred meters. It is a uniform gray, including the areas around craters and those within the bright albedo features, indicating the composition is of a carbonaceous chondritic material.   (NASA/JPL)

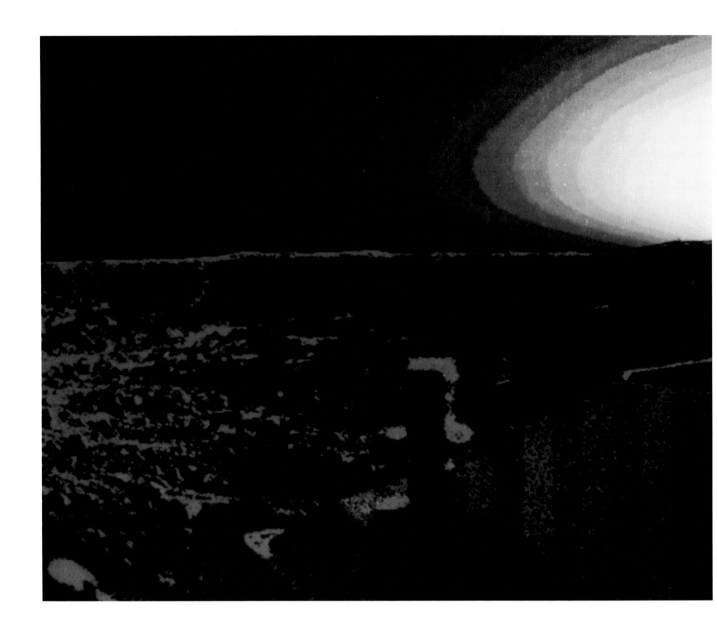

143. The early Mariner photographs were impressive—particularly those of Mariner 9—but Viking was able to send back spectacular images such as this striking Martian sunset. (NASA/JPL)

especially in the case of striking full-color images and the high resolution of selected surface features.

There was such a wild variety of terrain that Mars left no doubt that it was still evolving and changing. Winds were driving the erosion processes and causing major changes in the physical surface. Many areas showed the effects of severe flooding and volcanic activity from 50 million years ago, perhaps even more ancient than that. Where the floodwaters originated, and what happened to them, would remain a puzzle until the Viking lander arrived at the north polar cap.

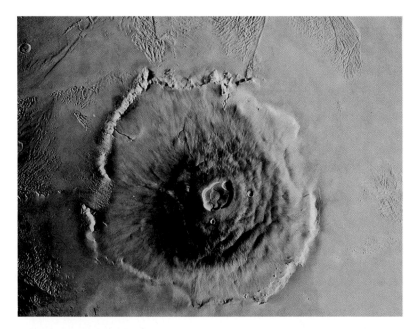

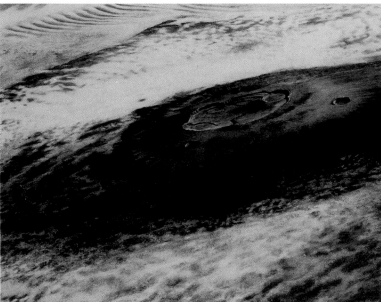

144, 145. These two photographs show two quite revealing angles of the great Martian volcano, Olympus Mons. On the top is the typical dead-on view of the fifty-mile crater. On the bottom, the fifteen-mile-high mountain as seen at mid-morning, encircled by clouds that reach up its flanks to an altitude of twelve miles, leaving the crater free of clouds.

(NASA/JPL)

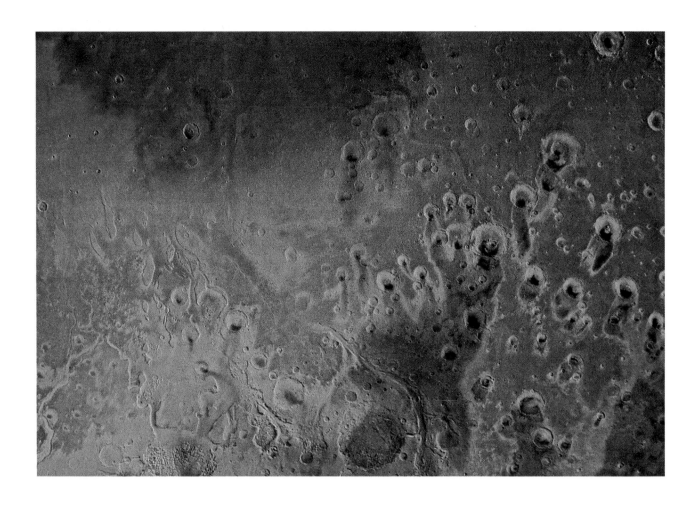

146. In this digital-image mosaic, water appears to have flowed over the surface, diverging around the craters and low hills, leaving a long tail downstream. The Oxia Palus region with its torturous patterns is similar to Earth terrain that is formed by catastrophic floods, wind erosion, and glacial action.

(U.S. Geological Survey, Flagstaff)

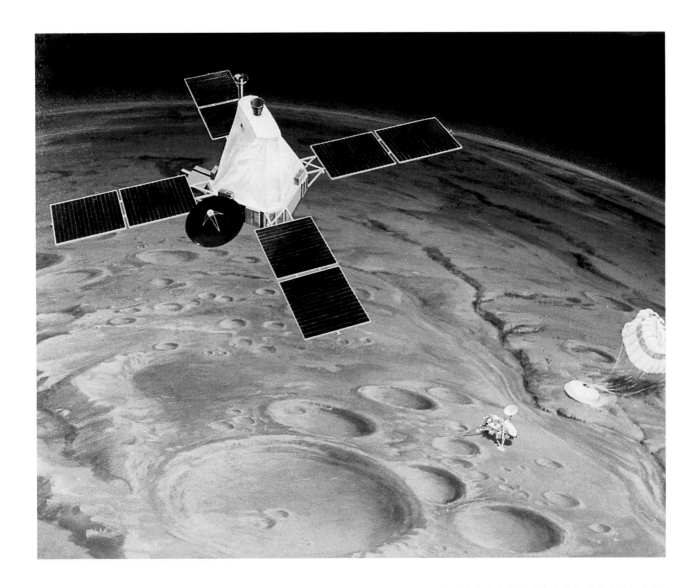

VIKING LANDERS

The age-old mystery of Mars, associated with war, monsters, and great civilizations, was finally breached when the Viking landers were sent to the surface. Nearly a year after the launch date, on July 20, 1976, Viking 1's lander touched down on the Martian Chryse Planitia (Plains of Gold). A month later, Viking 2 set down on Utopia Planitia.

The landers sent the orbiters a stream of horizontal landscape panoramas, stretching from the landing gear to the horizon. The longevity of the mission meant that scientists could compare views during every season of the Martian year. The landers sent

147. This artist's rendition shows the launch of a lander from the Viking orbiter. Ten thousand specialists labored for eight years to achieve this monumental step in space exploration. After the numerous Soviet failures, the U.S. team played it safe when they saw the rough surface at the proposed landing site. They aborted the first landing attempt, which had been intended to coincide with America's bicentennial celebration on July 4, 1976.   (NASA/JPL)

148. When Viking scientists got their first look at the location that had been selected as a safe, smooth, and hopefully scientifically interesting region to land Viking 1, Hal Mazursky of the U.S. Geological Survey cried out, "My God, it looks like the Bad Lands of South Dakota!"
(NASA/JPL)

back more than 4,500 pictures in all, while the orbiters added an astounding total of over 52,000 photographs.

First, scientists got their initial look at the true color of Mars. The view from Viking 1 clearly showed the spacecraft sitting under a salmon-pink sky. On the ground covered with boulders

149. As seen in this artist's rendition, the long-awaited moment arrived as the Viking 1 lander used its control rockets to lower itself to a gentle touchdown on the Martian surface. About the size of a Volkswagen, Viking was the first spacecraft to ever make a successful landing on Mars.
(NASA/JPL)

and fragmented rocks, the Martian soil appeared definitely red in color.

The first day on Mars for Viking came to a swift end in the thin air. By now the meteorological station had been erected and was working full time. Then the first Martian weather report reached Earth: "Light winds from the east in the late afternoon, changing to light winds from the southwest after midnight. Maximum wind, fifteen miles per hour. Temperature ranging from minus 122 degrees Fahrenheit just after dawn to minus 22 degrees Fahrenheit."

THE SEARCH FOR LIFE

Both the Viking landers dug trenches and extended magnets on a mechanical arm to capture soil samples. The soil clung to the magnets indicating there was strong contact.

With chemical analysis, iron and silicon were clearly

150. The images that appeared on the screens of Mission Control were monitored and processed by three huge computer systems. The sky was determined to be pink due to the scattering and reflection from reddish dust particles suspended in the lower atmosphere. Chemical analysis revealed that the sand is granular and dense, something like moist sand on Earth. (NASA/JPL)

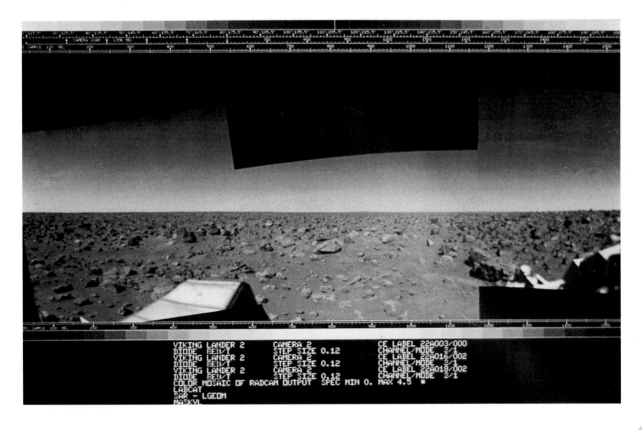

151. This is the first picture ever taken on the surface of another planet. The camera is about five feet above the footpad of the lander. In the immediate area, there are small rocks as well as finely granulated material, sand or dust. Many of the small rocks are flat, while several larger rocks have irregular surfaces with pits and cracks.
(NASA/JPL)

confirmed in the rich Martian soil. There was also an abundance of magnesium and almost a hundred times as much sulfur present than exists in either terrestrial or lunar soil.

The single most significant discovery of the Viking mission was the discovery of frozen water beneath a permafrost layer of soil. The north polar cap consisted of frozen *water* — not frozen carbon dioxide, as nearly every scientist had come to believe. They were amazed that abundant ice remained under the conditions of such a thin atmosphere. Its presence indicated that huge quantities of frozen water were beneath the surface.

This exciting discovery unleashed a flood of speculation that life would be found on Mars.

Gerald Soffen, the chief Viking project scientist based at the Langley Research Center in Hampton, Virginia, was the driving force behind the lander program to search for and hopefully

152. Taken about fifteen minutes before sunset, the low angle of the Sun finally highlights the variations of the terrain in this photo. There is a depression near the center, and toward the horizon, a distance of about two miles, several bright patches of bare bedrock can be clearly seen. (NASA/JPL)

confirm beyond all question or doubt that there were living organisms on Mars.

Soffen conducted the first experimental efforts to see if a tracer of carbon dioxide could be converted to the tissues of any microscopic Mars plant or bacteria. Initially there was a "marginal positive" result, yet subsequent weeks of repeated experiments all came up negative.

While the biology and organic chemistry experiments discovered chemical activity in surface samples, one major research team reported, "The mission uncovered no evidence of organic chemicals and/or other clear evidence that life has ever existed at the two Viking landing sites."

The situation was similar to that of the first Mars mission, Mariner 4, in which only 1 percent of the surface had been photographed. It was like trying to identify an elephant by feeling its trunk.

Soil samples needed to be gathered from more than two places, and they needed to be transported back to Earth for a thorough analysis. As Soffen stated, "One handful of dust tested in a decent laboratory for an afternoon would probably give us a final resolution."

153. On the left (opposite, top) is a small dune of Martian material left by the trenches that Viking 1's sampler scoop had dug. The scoop and its arm are seen here in the parked position.
(NASA/JPL)

· ·

154. As the Sun rose over Viking 1 sitting on the Martian surface on July 26, 1976, it illuminated the lander's American flag and bicentennial symbol (opposite, bottom).
(NASA/JPL)

· ·

155. The Viking orbiters captured this view of a local dust storm in the Argyre basin. The turbulent orange dust (see arrow) is roughly 200 miles across. With chemical analysis, geologists determined that the red color of the air and soil was created by a chemical reaction between iron and water (perhaps as water vapor in the atmosphere). This is the process known as "rusting."
(NASA/JPL)

· ·

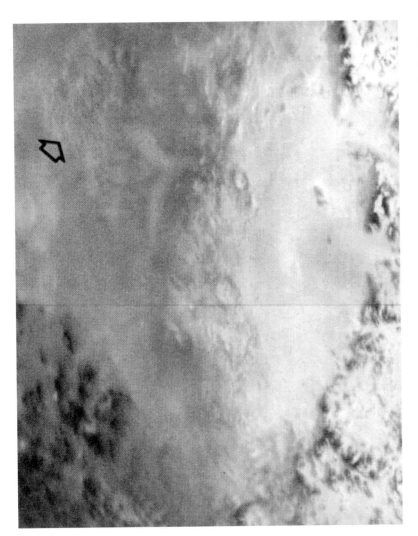

Chapter 11

MARS TODAY

It is likely that mankind will never give up the quest for a planet that is like our own, and we will continue to search for intelligent life on other worlds. When our first look around Mars for even a shred of life proved futile, extraterrestrial devotees turned their eyes beyond our own solar system.

With the refinement of radio astronomy in the latter half of the twentieth century, there was a dramatic increase in the possibility of picking up signals from an alien intelligence. Currently there are numerous projects connected with Search for Extra-Terrestrial Intelligence (SETI) that have been proposed or are underway. Most projects conduct spectral and E-M band wave observation of cosmic phenomenon many light years away from Earth.

The belief in the existence of alien life has, of course, led to a number of "contact stories." Whitley Strieber's alien abduction manifesto, *Communion* (1989), was supposedly based on the

Valles Marineris Region
60° Longitude

Tharsis Region
160° Longitude

Syrtis Major Region
270° Longitude

156. Mars has become cloudier in recent years, which means the planet is cooler and drier because water vapor in the atmosphere freezes out to form ice-crystal clouds. The Hubble Space Telescope, which created these color composites of Mars at opposition in February 1995, resolves features on Mars nearly as well as the Martian space probes. (NASA)

. . . .

author's personal experience. Others were based more firmly on SETI research, such as Carl Sagan's fictional rendition of an alien encounter in *Contact* (1984). *Contact* is best known for its depictions of the way scientists think, and for the intriguing theory of travel through the universe via black holes. (It is also noted for the two-million-dollar advance Sagan received in 1981, three years before *Contact* was written.)

Carl Sagan was one in a long line of astronomers who have written fictional accounts inspired by their observation and research. Sagan had always had a fond eye on Mars; he played an active role in the Mars experiments carried out by Mariner 9 (1971), and he also worked on the Viking and Voyager projects.

In his book *Cosmos*, published in 1980, Sagan asked, "Why so many eager speculations and ardent fantasies about Martians, rather than, say, Saturnians or Plutonians?" Sagan believed it was because of the surface similarities between Earth and Mars, such as the polar ice caps, the twenty-four-hour day, and even

157. At a Viking press conference, Carl Sagan is seen on the right, along with Gerald Soffen, Viking project scientist. Sagan was the director of the Laboratory for Planetary Studies at Cornell University, as well as the co-founder and president of the Planetary Society, a very large space-interest group. He was also responsible for placing a message to alien life aboard the interstellar spaceship Pioneer 10 (Jupiter flyby, 1973).

(Gerald Soffen)

the seasonal changes in the patterns that appear on the surface. And with the Viking data, there was new evidence of atmospheric clouds, surface water, and vast weather patterns.

As Sagan pointed out, it has always been tempting to think of Mars as a world filled with life. It was easy to imagine the red planet as the role model for our own cosmic struggles—and for our hopes of a highly advanced alien civilization that could show humanity the way to peace and prosperity.

158. Nearly one hundred years ago, Stanley Wood illustrated *Stories of Other Worlds* (1899), including this drawing of a conversation with a Martian. If you shrank this bulbous-headed, three-meter alien down to three feet in height, he would be a dead ringer for those pointed-chin abductionists that Whitley Strieber has helped make so famous. (Mary Evans Picture Library)

As a scientist, Sagan conducted experiments regarding the possibility of life on Mars. In the years before Viking, he and his colleagues created "Mars jars," chambers that could simulate the Martian environment. This was before the Viking landers recorded temperatures ranging from slightly above freezing to about -80 degrees Celsius prior to dawn.

Inside the Mars jars, the atmosphere was mostly carbon dioxide and nitrogen, while ultraviolet lamps reproduced the Sun's effect through the thin air. No standing water was present. Then Sagan stocked the jars with microbes, many of which froze to death after the first plunge in temperature. Others died of thirst or lack of oxygen. However, there were some that survived. "In other experiments, when small quantities of liquid water were present, the microbes actually grew," Sagan stated in *Cosmos*.

No matter how many experiments are done, the scientific world agrees that the only way to know for certain whether there is life on Mars is to go there.

ALIENS ON MARS

Though Sagan's novel *Contact* depicted aliens as beneficent to mankind, most fictional accounts accept Whitley Strieber's UFO abduction theory that aliens are malevolent in regard to humans. This paranoid twist on the old invasion fears actually has our bodies being probed, tested, and invaded—often with the memory of such visits wiped from the victim's mind.

Television was quick to speculate about aliens visiting Earth, in series such as *Project UFO* (1978–79) and the on-going television success of *The X-Files* (1993 debut). With the year 2000 approaching, there may well be an extra hysterical edge brought to the current fascination with mystics and promoters of the paranormal.

As always, paranoid fears rise in periods of anxiety and uncertainty, reviving the gothic belief

159. This fresco was found among others dated to 6000 B.C. in Tassili-n-ajjer in the central Sahara. Henri Lhote, a French archeologist, called this figure the "great Martian god." Though there is no evidence to support this title, many writers insist that this is a depiction of extraterrestrials who have visited Earth.   (Courtesy of the author)

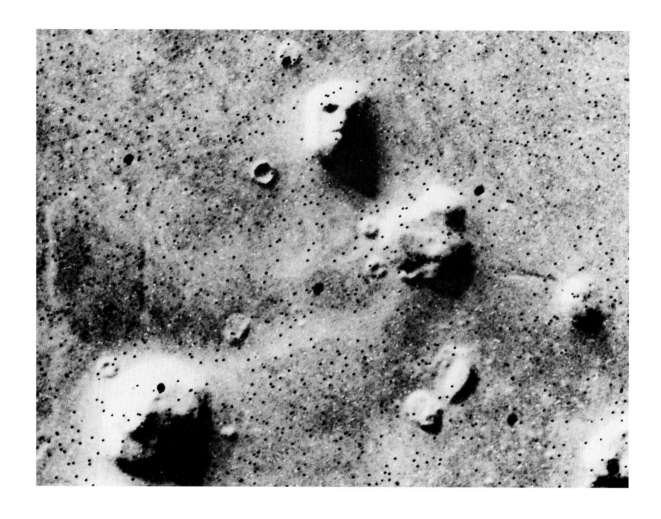

160. This mile-wide helmeted "face" on the Plains of Cydonia was photographed on July 25, 1976, by the Viking 1 Orbiter at a distance of 1,162 miles. Planetary geologists believe that the illusion of a face was created by the natural erosion of the rock. Others (including some trained scientists) claim that the Cydonia Face is a great stone monument, built by a long-departed and superior alien race. (NASA/JPL)

· ·

161. Here, in the picture on the television monitor, Gentry Lee points out "chicken tracks" across the surface to Gerald Soffen. The trail of depressions in the high-resolution image turned out to be markings on the surface made by pebbles blown around during landing by the thrust of the Viking lander's descent engines.   (Hans-Peter Biemann)

that we are not safe and that the universe contains menaces that can swoop down on us at any moment. American recent geopolitical history underwent a sharp escalation of the Cold War, culminating in President Ronald Reagan's proposed construction of the ultimate space defense system—Star Wars.

162. Is that sagebrush on the horizon? That's what scientists in the Viking Mission Control were asking when the Viking 1 lander took this dramatic view southeast over the Martian plain. What at first appeared to be vegetation was, after closer examination, a zone of large boulders on the distant horizon. Photos such as these prove that Mars is not all that it may seem to be.   (NASA/JPL)

But now, with the fall of the Berlin Wall and the collapse of the Soviet empire, the United States stands as the sole superpower on Earth. There continues to be a vociferous advocacy of the Star Wars system from elements within the government and the military industrial complex. What are they afraid of?

Under this "War of the Worlds" mentality, the myth of Mars has completed another abrupt reversal in the post-Viking decades. At first, direct contact with the Martian soil and indisputable photographs had led to a realistic trend in fiction, with writers attempt-

163. This famous Viking 1 photograph surprised scientists and laypersons alike. On the extreme left, above the housing of the lander's sampler scoop, is either the letter "B" or "S" etched on the surface of a large rock. Was it put there by a Martian? Hardly. Investigators decided it was an illusion caused by weathering and the angle of the Sun. (NASA/JPL)

ing to understand the reality of existence through the reality of a hostile Mars. Then in the early eighties, people began reverting to age-old conspiracy cants, insisting that there must be evidence of life on Mars, only the government is hiding it from us.

THE FACE OF CYDONIA

The photograph that started it all was taken while Viking 1 was searching for an alternate landing site for the Viking 2 lander. Picture Number 35A72 was taken of the plains of Cydonia in the northern latitudes of Mars. It clearly reveals the image of a humanoid face, approximately one mile in diameter.

The image is somewhat speckled in appearance; NASA explains this is due to missing data, called bit errors, lost while transmitting photographic data from Mars to Earth. Bit errors constitute part of one of the "eyes" and "nostrils" on the eroded rock that resembles a human face.

Gerald Soffen may have unwittingly started a paranoid reaction when he told a reporter friend (with tongue firmly in cheek) that he had seen "the first actual photo of a Martian."

Planetary geologists attribute the origin of the face formation to purely natural processes, pointing to the shadows in the rocks that give the illusion of a nose and mouth. NASA was not quick to leap to conclusions; the scientists had been well trained by Mariner missions to be cautious when it came to viewing Mars and rendering "expert" opinions. After all, several Mariner missions failed to detect the largest volcanoes in all the solar system. And with every mission and each photo taken, more details continue to emerge. The 1997 Surveyor and Pathfinder missions should provide even more data that will undoubtedly shatter older, well-cherished beliefs about Mars.

The discovery of the face on the plains of Cydonia swiftly mushroomed into a cause célèbre by hundreds of writers and alleged experts who were soon castigating NASA for holding back information. They accused NASA of being afraid that the American public would panic if the truth was admitted— that the face was an artificial artifact left by a long-departed species of superior intelligence.

Some, like Richard C. Hoagland, fanned the flames of a government cover-up conspiracy. Hoagland reasoned that the face had been planned and built with significant messages for those first to encounter the monument staring upward from Mars. He even

164. The U.S. Geological Survey processed about a thousand Viking photographs to fully chart Mars in a series of digital-image mosaics. This image captures the Mare Boreum region, with the central part covered by a residual ice cap. Spiral-patterned troughs expose the layered terrain beneath, and the cap is surrounded by broad, flat plains and large dune fields, resembling Earth-like terrain. (U.S. Geological Survey/NASA)

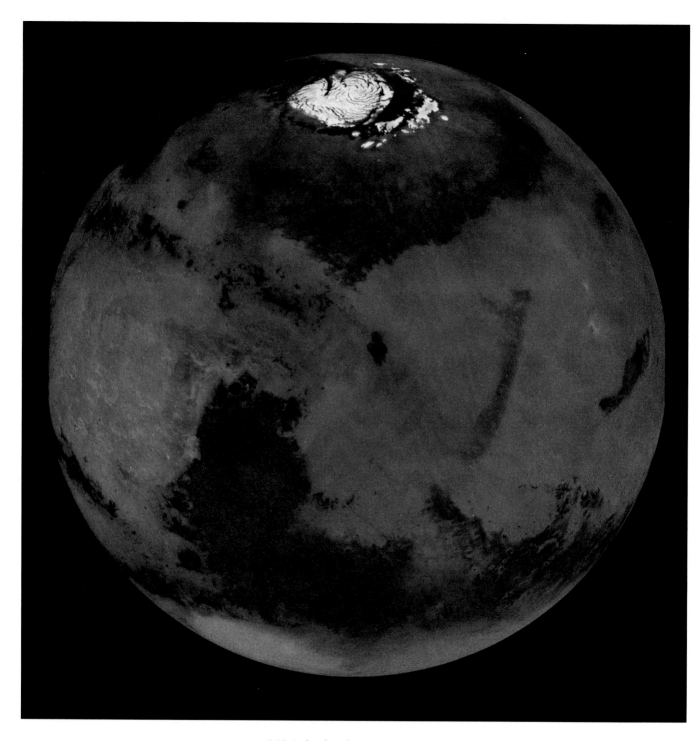

165. In this digital mosaic we can see an enormous band of sand dunes encircling the polar cap. They may prove to be the largest dune fields in the solar system. Geologists say the dunes could have formed from dust settling into the ice of the cap, which has alternately melted and grown with the Martian seasons for hundreds of millions of years. (U.S. Geological Survey/NASA)

reasoned that the Cydonia Face was never intended to reveal what the alien builders looked like, but it predicted what man would look like. A very neat trick for a people who were on Mars perhaps a million years before humans existed and who left us this monster edifice as a parting gift.

A superb effort to advance a theory of the face as a relic of an intelligent life-form soon appeared in bookstores with the quiet and self-effacing title of *The McDaniel Report* (1994). The writing was quite persuasive as to the truth of the Cydonia Face as an extraterrestrial creation, and was backed by Stanley V. McDaniel's powerful credentials. He doggedly put all the information together—research, data breakdown, extrapolation, and photographs of Mars and Earth monuments. In his book he leveled the accusation that the U.S. government, and NASA in particular, knew full well how the face came to be but refused to share that knowledge with the American public and the rest of the world.

People are willing to believe McDaniel's theory, because quite often in the (recent) past, U.S. government agencies have been shown to have misrepresented, distorted, and lied outrageously to the public on many important matters—from Watergate to the Iran-Contra affair.

Recently a massive marketing campaign has sprung

166. Mars Observer is shown in this General Electric artist's concept as the spacecraft nears the red planet. But after months of interplanetary travel, just days away from photographing the surface in detail, Observer became a dead piece of high-tech junk. NASA scientists haven't heard from the probe since, and they are unsure of what exactly caused this mission to fail.

(GE Aerospace Art)

up, with space-oriented clubs and organizations charging membership fees and selling all sorts of associated products — videos, games, books, clothing, and trinkets that support the basic tenet that there is life on other worlds and that it is among us.

These groups have differing purposes and perspectives. The Mars Project asks for money to support an investigation of the Cydonia Face, imploring "we are in a race against time." At the opposite end of the scale, the American Astronautical Society works hands-on with government space programs.

MARS OBSERVER

Mars Observer was scheduled for a 1988 launch from Cape Canaveral, but it did not lift off until 1992. Along with the delay, the probe was increasingly loaded with mythical baggage and the curses of thousands who were eager for Observer to finally prove that the American government was "holding out on the people."

One of the missions of Mars Observer, the most advanced and expensive U.S. spacecraft to be sent to Mars, was to photograph the Cydonia Face in detail. Observer was also supposed to locate landing sites for American astronauts who would (hopefully) arrive sometime in the near future.

Three days before Observer was moved into position around Mars—after months of successful space travel to reach the planet—it turned into an enormous piece of dead junk. The same thing had happened to *seventeen* earlier Russian missions, defying all odds. Despite the brightly shining mission results of the Mariner and Viking projects, NASA's very real fears of failure were confirmed. Once again, budget cuts immediately followed Observer's demise.

Of course, in the spirit of the Cydonia Face itself, some people believed that *something* on or near Mars had taken aim at the spacecraft and destroyed its electronics.

Then the paranoia shifted in a familiar pattern, and the blame was pinned on both NASA and Russian scientists. They were accused of cutting off all contact with their probes because of what they found. Apparently it was something so horrible, so frightening, that politicians and civil servants believed the sensibilities of the public were in dire need of protection.

The tabloids were filled with reports that Mars Observer was still working, sending back to Earth all kinds of secret photos and reports meant only for the eyes of top government officials.

What is the real answer to why these probes failed? Easy. They are highly complex pieces of machinery and electronics, and as such they are subject to failure.

Even Earth-orbiting probes have been notoriously difficult to put into operation. Shortly after the Hubble Space Telescope was launched in April 1990, NASA had to admit that many of their most essential scientific instruments were not working. Even the eight-foot primary mirror (which required five years of grinding and polishing) was sending back pictures that were fuzzy.

167. Hollywood has always been big on remakes, but in the past couple of decades there has been a resurgence of sci-fi remakes, some good, like *Invasion of the Body Snatchers* (1978) and others that are pretty bad, like *Invaders from Mars* (1986). This still from *Invaders from Mars* reveals the melodrama that was typical of the poorer remakes, also seen in *Buck Rogers in the Twenty-fifth Century* (1979) and *Flash Gordon* (1980). (Everett Collection)

A massive campaign was launched to "fix" the mirror so it could be focused. The space shuttle program became the savior of the Hubble Telescope when astronauts went into orbit in 1993 and successfully replaced Hubble's solar panels, magnetometers, gyroscopes, relays for the spectrograph, and, most important, the computer.

NASA does an outstanding job of dealing with delicate hardware that must withstand the massive gravitational forces which take place during a launch—but it has never been easy. The first nineteen lunar probes failed to complete their assigned missions. The fourth Mars launch, Mariner 8, did not even get out of Earth's atmosphere. The public does not need to look much further than wiring and microchips to find reasons for why Mars is so difficult to reach.

168. In *Total Recall* (1990), the protagonists are saved by enormous nuclear reactors that are buried deep beneath the ground. In the role of a futuristic Rambo, Arnold Schwarzenegger gains access to these machines, which release an explosive cloud of life-saving gas. In the grandest tradition of space opera, mankind can now make a new home in the transformed Martian ecology.
(Everett Collection)

FICTIONAL MARS

As always in science fiction, there was a developmental leap based on the new scientific knowledge that had been gathered about Mars from the Viking missions and the most recent opti-

cal surveys completed by the Hubble Space Telescope. With the surface clearly an inhospitable place, writers began depicting life on Mars through alternative realities.

Harry Turtledove's *A World of Difference* (1989) described an alternate Mars in the form of a sword and sorcery fantasy. Mars was called Minerva, and the Minervans battled the arrival of two manned missions, one from the United States and the other from the Soviets. (This kind of Cold War scenario is fading into history.)

Mixing genres has become one of the newest trends, as in John E. Stith's *Death Tolls* (1987), which used the setting of a terraformed Mars for a hard-boiled detective mystery. Interstellar bounty hunters were found in William C. Dietz's *Mars Prime* (1992), while Jack Butler created an effective presentation of human settlements on Mars in *Nightshade* (1989), including a scientific rationale for vampires.

SATIRE

For many serious sci-fi writers, Mars is only fit for satire. The reality of Mars has made novels such as Theron Raines's *The Singing: A Fable about What Makes Us Human* (1988) seem quaint and outmoded. Raines had a Martian spacecraft crashing into the Guggenheim Museum in New York, where one of the Martians, according to plan, met and impregnated a human girl. For most people, this straightforward rendition is a little hard to swallow.

Terry Bisson, on the other hand, used the very real impetus of making money to spur a journey to Mars as part of a movie-making scheme in *Voyage to the Red Planet* (1990). Bisson's portrayal of aging would-be colonists traveling in the decrepit umbrella-shaped spaceship called the *Mary Poppins* was both scathing in its view of modern mercenary ideals and haunting in its poignant questioning of what might have been.

Frederick Pohl also wrote a light-hearted, romantic story of a confrontation between humans and aliens in *The Day the Martians Came* (1988). More recently, Mac Wellman's *Annie Salem* (1996) described a fantastical quest taken by the protagonist in order to win the woman he loves. He traveled to Mars and met a strange brew of creatures, accomplishing patently heroic feats before undergoing a process of "pumkinification." He eventually returned to his hometown in the Midwest with nothing more than an altered perception of reality.

CYBERPUNK

Unlike fantasy, hard sci-fi developed alternate realities in a new style called cyberpunk. The cyberpunk plot is usually based on techno-networks of both real and artificial intelligences. Whereas the staple of sci-fi politics used to be international competition, in cyberpunk the power structures span the globe, united by information networks. The "punks," or protagonists, are antiestablishment heroes who chip away at the layers of illusion that blind the general public.

One of the formative pre-cynberpunk writers, Ian Watson, was concerned with the nature of perception in thought and language. Often his plots involved the control of information from the government to the people. In *The Martian Inca* (1977), Watson explored the possibility that a transformative virus existed on Mars, capable of adapting mankind to life on that planet. This story was published the same year that Viking landed and scooped up some of the rusty soil.

Like most cyberpunk writers, Watson examined the cost of achieving altered perceptions, especially those induced by outside agents like drugs. In *Martian Inca*, the remnant of alien intelligences that linger in the body were the price the colonists had to pay for their transformation.

A similar type of subjective conflict was explored in Philip K. Dick's *Do Androids Dream of Electric Sheep?* (1968). Dick's story became the basis of the most influential sci-fi film in recent decades—the superbly crafted *Blade Runner* (1982). Ridley Scott's hi-tech version of Dick's story rendered it into cyberpunk, and rather than the animal empathy story line, the key issue was a replicant's delusion that she was human. The bounty hunter eventually escaped with her rather than "retire" his beloved.

Another one of Dick's stories, "We Can Remember It for You

169. This color composite was made by the Hubble Space Telescope in December 1990. The image is blurry, as were all the Hubble photographs before 1993, yet it reveals more detail than Earth-based telescopes. Movies such as *Total Recall* really could not be blamed for envisioning Mars, with its thick canopy of bluish clouds and icy north polar regions, as a place where Martians once thrived.   (NASA)

· ·

Wholesale" (1966), became the basis of a 1990 film, *Total Recall*. With Arnold Schwarzenegger in the lead, *Total Recall* typified the contemporary attitude toward outer space. By turning the original into an action-packed, psychological thriller that never let up its driving pace, a box office-busting winner was created. It also did not hurt to have a purported $60-million budget, making it at the time one of the most expensive films ever made. Now it barely keeps company with even more expensive sci-fi films such as *Terminator 2* (1991) and *Waterworld* (1995).

In Dick's original story a man attempted to purchase false memories of a trip to Mars, only to discover his real memories had been suppressed. In the screen version, the false memories looked the same as the true ones, making it difficult to understand what was illusion and what was reality. The hero had to fight the villains' psychological control while performing Edisonade feats of daring.

Though the production was quite realistic, making the environment remarkably Mars-like, *Total Recall* was jarring in its inconsistencies. There was no difference in gravity between Earth and Mars, even though a 200-pound man on Earth would weight only 76 pounds on Mars. And the glass frames of the Martian structures were hopelessly weak, prone to breaking into thousands of pieces with little more than a hard shove. The general mayhem when the dome does shatter culminated in awful eye-popping and gurgling effects.

The viewer must also get past the litter of ideas and plot devices that had been borrowed from a wide range of sources. Most telling was the gratuitous violence that has become typical as Hollywood tries each year to top the previous year's spectacular productions. Unfortunately, *Total Recall* was studded with fighting, shooting, and people being burned alive; almost everything in sight was destroyed. Oddly enough, this violence was combined with a flippant, larger-than-life tone that reflected comic-book adventure, and could be seen in other contemporary films, such as *The Mask* (1994).

Total Recall also relied rather heavily on the fantasy element of ancient Martian machines that can release billions of tons of oxygen into the thin atmosphere. This air-making machine was straight out of Edgar Rice Burroughs's *A Princess of Mars* (1917), and, like all good space opera devices, it satisfied the deep-rooted belief of the audience that at one time Mars did have a thriving and highly advanced civilization, and that the red planet could once more be made habitable.

Chapter 12

MARS TOMORROW:
THE DAWN OF A NEW MILLENNIUM

The prospect that life once existed on Mars is being raised following analysis of a meteorite recovered on Earth." This excerpt from a brief article in the weekly aerospace newspaper *Space News* on August 5, 1996, opened the floodgates of speculation about life on Mars.

Indications of single-cell bacterial life-forms were found imbedded in a meteorite dislodged from Mars millions of years ago. The meteorite landed in Antarctica some thirteen thousand years ago and was discovered in 1984.

The preliminary research on the clay inside the meteorite from Mars was finished in July 1996. Though their results are "not definitive," NASA scientists claim the microscopic tubular structures are fossilized bacterial forms. Both electron microscopes and laser mass spectrometry were used to compare the microbial forms of Martian life with those of known microfossils on Earth.

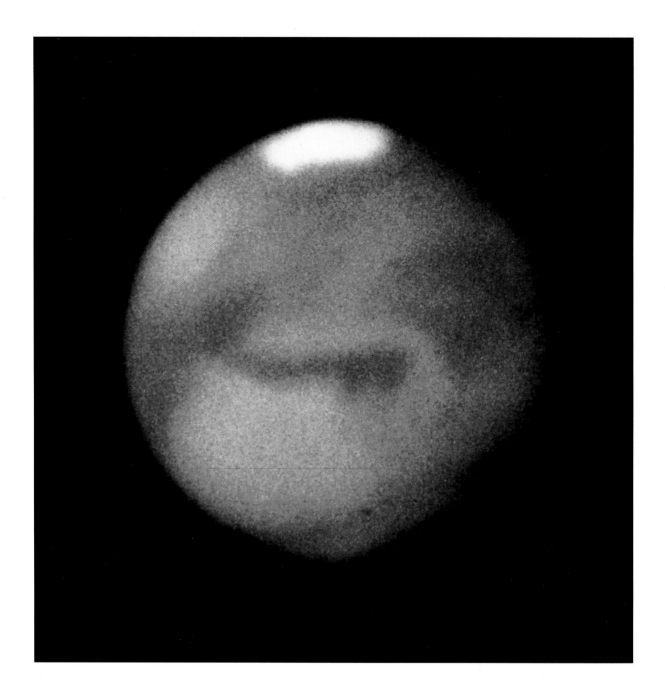

170. Much of the twentieth century has been dominated by mankind's fascination with the possibility of life on Mars. In the early half of the century, it was Martians—either invading or offering advice—that quickened the public's imagination. Now as the world prepares to enter the twenty-first century, man turns his eyes to smaller prey, hoping to find hard evidence that life is possible beyond our own planet.   (NASA)

ALH84001,0

171. This is an end-view image of the Martian meteorite that was dislodged from Mars about sixteen million years ago. The meteorite was found in the Allan Hills ice field in Antarctica by an annual expedition of the National Science Foundation's Antarctic Meteorite Program; in 1984. It has been preserved for study at the Johnson Space Center's Meteorite Processing Laboratory in Houston, Texas. (NASA)

David S. McKay, the geochemist who led the NASA investigating team at the Johnson Space Center in Houston, reports they also found an abundance of calcium carbonate in the meteorite, material that is sometimes associated with life processes.

Some planetary scientists disagree with McKay. Ralph Harvey from Case Western Reserve University in Cleveland and Harry Y. McSween from the University of Tennessee at Knoxville point to their previous studies reported in the July 1996 issue of *Nature*, in which they found no water-bearing minerals in the Allan Hills 84001 meteorite (as it has been officially designated).

Harvey and McSween began conducting new tests within weeks after preliminary findings from the NASA scientists were released. The two believe that the tubular structures were formed by the superheating effect (approximately 1,200 degrees Fahrenheit) caused by the asteroid impact that knocked the meteorite loose from the Martian soil.

Countering that theory is Everett K. Gibson Jr., at the Johnson Space Center. He says the presence of certain oxygen isotopes, minerals, and organic molecules

172. These are the orange-colored carbonate mineral globules found in the Allan Hills meteorite. Their structure and chemistry suggest that they may have been formed with the assistance of primitive, bacteria-like living organisms. (NASA)

MARS TOMORROW: THE DAWN OF A NEW MILLENNIUM

199

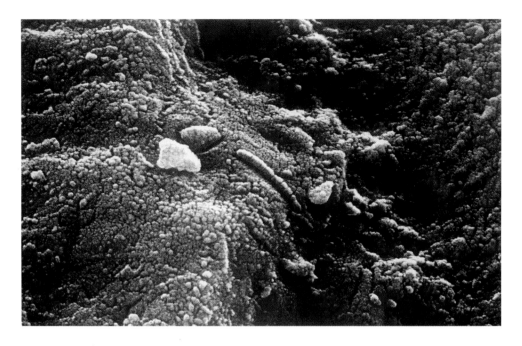

173. This high-resolution scanning electron microscope image shows an unusual tubelike form that is less than one-hundredth the width of a human hair. Although this structure is not part of the two-year NASA research effort published in the August 16, 1996, issue of the journal *Science*, it is located in a similar carbonate globule in the meteorite. This structure will be the subject of future investigations to confirm whether or not it is fossil evidence of primitive life on Mars.   (NASA)

indicate the carbonates formed at temperatures no higher than 180 degrees.

The microfossils—if that is what they are—are many times smaller than the diameter of a human hair. Scientists must now look inside these specks to try to determine if there are cell walls inside. NASA is planning to use an ultrasensitive transmission electron microscope to examine thin slices of the meteorite, hoping to magnify the fossils. They will be looking for anything resembling chlorophyll or DNA proteins, the building blocks of life as we know it.

The carbonate deposits found in the meteorite are thought to be 3.6 billion years old. According to available data, scientists believe that the Martian crust developed 4.5 billion years ago, soon after the planet formed. Asteroid impacts could have caused surface fracturing in the crust, creating crevices where carbonate globules would support bacterial life.

With evidence from orbital photographs taken by both the Viking mission and Mariner 9, scientists are convinced that water once flowed over the Martian landscape, leaving behind channels, lake beds, and river deltas. It is possible that life once existed—and may still exist—on Mars. The trick is, how do we find out?

Based on the fossils found in the Mars meteorite, in November 1996 President Clinton called for a summit involving scientists, national leaders, and international space experts to "plan a strategy." At this summit, the National Research Council of the National Academy of Sciences will present a report completed in mid-1996 stating that the current program for Mars exploration is "severely underfinanced."

Aside from the budget wrangling, the strategy for examining Mars has already been mapped out. The possible discovery of fossilized life has merely given a new sense of urgency to the projects already underway to search for life on Mars.

The latest Mars push began in 1993 with a major international meeting known as the Mars Summit, which inspired the first cooperative attempts to access, research, study, under-

174. The first American spacecraft to be launched was NASA's Global Surveyor. We see in this artist's conception how it will circle the planet in order to construct extremely detailed maps to be used by the worldwide scientific community. Surveyor's extended mission will also allow for detailed study of the climate on Mars.
(NASA/Lockheed Martin Astronautics)

MARS TOMORROW: THE DAWN OF A NEW MILLENNIUM

stand, and occupy the fourth planet, turning it into another home for the human race. The first summit, assembled in Germany, brought together scientists from NASA, the European Space Agency (ESA), Russia, France, and Japan. These scientists agreed that the goal of their cooperation would be to obtain as broad a field of research as possible in order to establish successful round-trip visits to Mars.

Russia launched a large Mars orbiter on November 17, 1996, but sadly, like so many previous Russian attempts to explore Mars, the mission failed. The space probe, originally planned for launch in 1992, had been delayed by Russia's economic and

175. This artist's drawing captures the descent of the Pathfinder through the Martian atmosphere. The sequence shows, starting at the upper left, the planetary ship decoupling from a cruise stage, then heating up as it enters the Martian atmosphere. Shortly thereafter, a parachute is fully inflated some five to six miles above the surface of Mars and the heat shield is released. (NASA)

political problems. The orbiter was to have surveyed Mars with European remote-sensing instruments, while two landers were to be sent to the surface. The landers were equipped with penetrators that were to have dug at least six feet into the frozen surface, probing the temperature and mineral composition of the soil. All came to naught, however, when the Mars craft could not break out of Earth's orbit and crashed into the Pacific Ocean.

NASA is also part of the first line of attack in the Mars initiative. Actually, NASA has ten missions planned for the next decade.

176. Here the Pathfinder lander is firmly on the surface, its panels opened to release the minirover, which is seen investigating a nearby rock. Project scientists predict the landing site is ideal for searching out life-forms that may still exist, or for finding evidence of life that died off millions of years ago. (NASA)

Just two weeks before the official results of indications of life on Mars were announced, NASA released a press statement regarding the launch of two spacecraft in November and December 1996, one of which is projected to reach Mars as early as July 1997.

The Mars Global Surveyor was the first to blast off from Cape Canaveral on November 7, 1996, but it won't reach Mars until September 1997. This $155-million project will circle the planet, map its surface geology, and study the climate of the thin carbon-dioxide atmosphere.

The second mission, Pathfinder, was launched on December 4, 1996, and will parachute to the Martian surface in July 1997. Pathfinder, consisting of a lander and a minirover, will land on the surface via parachutes and airbags, to begin soil and atmospheric testing. The $173-million project rests on the performance of a crawling robot the size of a little red wagon. NASA has learned to keep their programs restricted in size and cost after the Observer failure wiped out nearly a billion dollars.

Donna Shirley, director of NASA's Office of Mars Explorations, says that the landing site that has been chosen for Pathfinder "could not be better for scientists looking for water on Mars or possible life." The Ares Vallis site is at the mouth of an ancient water channel just north of the Mars equator. It will also provide access to rocks that have fallen from the headlands.

The twenty-two-pound roving vehicle, named Sojourner, will take short excursions beyond the lander to examine the chemistry of rocks with a spectrometer. The rover is powered by a larger robot, whose nuclear drive will keep the rover operating for some time. Major mission changes will be made through direct radio orders from Earth.

FUTURE MISSIONS TO MARS

Center National d'Études Spatiales, or CNES, the French space agency, is developing huge balloons to be released by their landers. (The scope of its program has expanded to include partnerships with the Soviets and the Americans.) Each balloon is projected to be 150 feet high, bobbing gently above the dusty

177. NASA intends for future Mars missions to include both orbiters and landers, as seen here in an artist's rendition of the proposed 1998 mission. (NASA/Lockheed Martin Astronautics)

surface, and propelled across huge distances of Mars rather than crawling about within the limited range of the ground rovers. The balloons will last up to several months, as long as they do not slam into jagged terrain or fall off significant cliffs. A "snake" will be suspended beneath each balloon by a 450-foot cable. The three-inch-wide aluminum snake will contain instruments and will float a few thousand feet above the ground in daylight. At night the balloons will descend closer to the surface, dragging the snakes along the ground, where their instruments will search for signs of water.

In 1998, NASA is planning to send two more spacecraft to Mars—again an orbiter and a lander. The orbiter will study the upper atmosphere, while the lander will examine the layers of dust and ice near the south pole. In addition, Donna Shirley

178. The moon Phobos is tiny compared to other moons in our solar system—about twice the size of a large mountain on Earth. Yet both its position and its speed of rotation around Mars make it a superb outpost for berthing spacecraft, for temporary storage of critical nuclear materials, and as a transfer point for cargo drops to the Martian surface. Phobos could also be the site of mining operations.   (NASA/Courtesy of Robert McCall)

states that the plans for 2001 are "wide open." Equipment could include a gamma-ray spectrometer to survey the global chemical composition of the surface of Mars. Infrared spectrometers could search for water repositories. The lander part of the mission could include the delivery of a comparatively large Russian roving vehicle to the surface.

NASA also has plans on the boards for a robotic return mission to gather Martian rocks and soil and bring them back to Earth for tests. This mission was tentatively set for 2005, but the intense interest ignited by the Mars meteorite has led NASA to push up the projected launch date to 2003.

A mission to fly to Mars and return with soil and air samples is a much more complicated and expensive program than Pathfinder or Global Surveyor. Yet it is the logical next step before sending human beings to Mars. Scientists hope that robotic probes can obtain the samples needed to discover whether there is life on the planet.

One bit of evidence that is clearly either bacterial or viral will forever change our concept of ourselves and our relationship with Mars. For if Mars once did harbor life, then we face the challenge of bringing life back to that planet again.

MANNED MISSIONS TO MARS

After the Mars meteorite findings, Congressman Dave Weldon, vice chairman of the House Space Subcommittee stated, "What makes sense for us to do is for us to start planning for a manned mission to Mars."

NASA administrator Daniel S. Goldin agrees; he has long proposed an international manned flight to the red planet, when Earth and Mars are closest and the travel time would be at a minimum. Yet Golding also insists that "Our missions should be driven by scientific potential and the potential for economic opportunity."

Currently, NASA is spending about $100 million a year on Mars probes and research. Based on the meteorite findings of

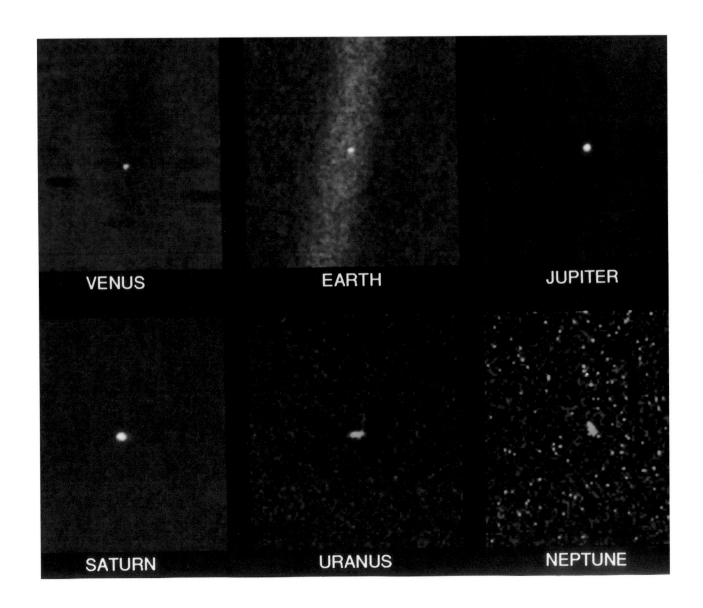

VENUS EARTH JUPITER

SATURN URANUS NEPTUNE

179. The first humans on Mars will see this view of the night sky. Taken by Voyager 1, the top middle panel shows Earth appearing just like any other star. Even in this fuzzy image, the elongated "double" shape can be seen, evidence of our large and brightly lit moon. (NASA/JPL)

180. This early-twentieth-century artist's rendition is fairly accurate in its prediction of the probable surface character of Mars. Interplanetary explorers will find a land with many features: craters, deserts, rills, dry riverbeds, plateaus, and enormously grand volcanoes. They will also find ancient lakes and oceans long since dried, the crust cracked in the harsh sunlight. (Mary Evans Picture Library)

· ·

bacterial fossils, Goldin is prepared to take the next step to find evidence of extraterrestrial life. But in the early 1990s, NASA determined that the cost of sending a manned expedition to Mars would be about $400 billion. Scientists recognize that government funding may not cover the bill.

The ongoing robotic probe program came out of a need to economize. Donna Shirley stated, "We're going to fly smaller, cheaper spacecraft more quickly so that if any one fails, it doesn't destroy the whole program."

The Observer failure cost the space program more than money; NASA lost years of development time when its best engineers and support people fled to better-paying jobs in the private sector. This started soon after Apollo (with the last three scheduled flights canceled) and persisted with the next major program, the Space Shuttle. Instead of going up and down, the ships went around in circles—not too much of an advancement, as far as the public was concerned. But the shuttle program certainly boosted satellite technology, mainly through private enterprise. This development, by the way, helped usher in the existence of the World-Wide Web.

So it is not surprising that private companies started nosing around the idea of exploring Mars. Robert M. Zubrin, owner of the research and development company Pioneer Astronautics in Denver, worked out the costs involved in sending a manned mission to Mars and got a much lower figure than NASA's— $5 billion.

In 1990, Zubrin suggested that the U.S. government fund the project by offering a prize of $20 billion for the first manned expedition to make it to Mars and back. At even four times the projected cost, private companies would vie for the chance. If someone actually made it to Mars and back, NASA would have to pay only one-twentieth of the projected cost of doing it with public funds. And if no one were successful, taxpayers would not be out a dime for the attempts.

This method of offering prizes to explorers is not new. The fifteenth-century Portuguese and Spanish rulers often put up cash rewards for those willing to risk the voyage across the At-

Mars Base

MSFC-11/85-PA 4000-592A

181. This artist's conception of the first base on Mars shows how it will be made safe for explorers. They will need everything from water to food grown in clear domes. Their suits and structures must be capable of protecting them against the low pressure and the lethal radiation from the Sun. (NASA)

lantic or around Cape Horn of South Africa. In June 1996, the X Prize Foundation announced a $10 million prize for the first manned, completely reusable spacecraft that could take off on its own, rocket sixty-two miles above Earth, and return safely to make another flight.

It seems sensible to use a solid foundation of capitalism to better advantage: let people who make it their job to get things done efficiently take us to Mars for a profit. Even science fiction has bowed to this inevitability, as in Ben Bova's *Empire Builders* (1993), in which entrepreneurs made much more headway in space than the foundering national governments. In another of Bova's earlier works, *Mars* (1992), he realistically described the first manned mission to the red planet.

Larry Nivens wrote the *Dream Park* sequence according to the cyberpunk view of the universe. His *Dream Park* (1981), *The Barsoom Project* (1989), and *Dream Park: The Voodoo Game* (1991), all written with Steven Barnes, described a "virtual reality" Mars in a futuristic adventure park, where the public paid to participate in complex role-playing games.

CORPORATE INCENTIVE

What is on Mars that makes all this effort worth the trouble? One of the weakest underpinnings of the Apollo Project was that it merely brought back a few hundred pounds of Moon rocks. These rocks were immediately locked up for security, with NASA handing out miserly samples for study.

The old sea voyagers set out in quest of better goods and better lives. The explorers brought back new foods, spices, textiles,

182. NASA has always toted out the flags and created photo-ops in order to generate interest in the U.S. space program. Critics, such as Rick Tomlinson, claim that there is an easier way, which will save billions: turn exploration over to the private sector who can make money out of missions to Mars—and bring in results far under NASA's budgets. (Everett Collection)

precious stones, and, above all, gold. At heart, these romantic quests were moneymaking ventures.

To be successful, a colony on Mars would need to return some sort of goods to Earth. As the Apollo missions proved, it costs too much to move spacecraft through space without providing some kind of return.

Selling souvenirs from Mars has always been a thriving business. The sale of meteorites rose sharply after the discovery of microfossils in the Allan Hills rock, yet there is a severely limited supply of meteorites on the open market. Most are owned by museums, who were right on the heels of the space-interest groups in publicizing their possession of meteorites after NASA's announcement, attempting to draw crowds to view their Martian rocks.

These pieces of the red planet are sometimes called tektites. Most are small, pitted rocks—the Allan Hills meteorite is the size of a potato. They are part of the debris knocked loose by asteroid explosions.

The majority of Martian tektites favor Earth's Southern Hemisphere as their final destination. Antarctica seems to harbor the mother lode of these collectors' items. Most are much younger than the Allan Hills meteorite, and scientists do not hold out too much hope for finding microbial life inside of them.

But rocks from Mars could harbor fossilized extraterrestrial life. The demand for these rare cosmic relics could cause a boom in the sale of Martian souvenirs. Just look what happened with Pet Rocks.

Yet transporting rocks is extremely expensive using current technology. Each rock that is brought from Mars will cost hundreds more than its weight in gold.

The one thing that is inexpensive to transport from Mars is photo transmissions. There has been no shortage of stunning photographs of Mars in different phases and seasons, even details such as Olympus Mons and the Chyrse Canyon. Orbital Surveyor and Pathfinder should return an even larger bonanza of striking, awe-inspiring images.

Dozens of companies advertising in space, astronomy, science, and aviation magazines, as well as NASA installations, sell these pictures. Yet the image can be worth more than simply its souvenir price. Dealing with images naturally leads to the media with its vast resources of capital in advertising and sponsorship.

MARKETING MARS

John Tierney recently published "How to Get to Mars (And Make Millions!)" in *The New York Times Magazine* (May 26, 1996). It was revealingly subtitled "A Not Entirely Far-Fetched Prospectus for a Private Expedition in Space."

Tierney suggests that the first manned expedition to Mars be turned into a moneymaking media event comparable to the Olympic Games. He suggests expeditions be sponsored by corporations that would pay millions to be associated with Mars. After all, Coca-Cola spent $40 million to be an Olympic sponsor in 1996, and that would go a long way toward getting to Mars.

Jay Coleman, chief executive of the E.M.C.I. marketing company and an expert in special-event promotions, says Mars has high associative power: "If you're AT&T, it's a great way to align your brand with high technology and the future." Companies would pay large sums simply to be associated with Mars. NASA could even sell the rights to name prominent areas of geography.

Brandon Steiner, owner of a leading sports marketing company, says he would have fun marketing "the first sneaker to touch down on Mars." Steiner would not be averse to putting a Nike logo on the first manned spaceship to Mars.

Joel Rosenman, one of the promoters of the original Woodstock festival, suggests that a global consciousness of Mars could be created, with companies marketing products such as "Mars cereal." Rosenman also points out that a manned mission would give you "a $5-billion set" for a drama or series of docudramas.

John Tierney agrees that companies could not lose, pointing out that explorer Brindisi Shackleton failed to reach the South Pole, but his mission returned to Europe a success simply on the basis of one magic sound bite for the media. After all, the most spectacular failure of the Apollo program, Apollo 13, was turned into the most compelling movie about NASA to date, *Apollo 13*.

Tom Zito, president of Digital Pictures, says, "Millions of kids will pay to play the game and start to think of the astronauts as their friends." Av Westin, the producer who developed *Close-up* and *20/20* at ABC says, "The networks won't want to pay for access to the mission. But once cable pays and the public gets interested, they'll find a way." Westin suggests that the astronauts could do a weekly news update show or documentaries.

John Tierney believes that the way to Mars is through the example of Antarctica (where, ironically, the Allan Hills meteorite was found). Exploring and charting Antarctica was a hazardous task that required sizeable developments in technology simply to keep the scientists alive.

Yet once polar explorers had established a beachhead, tourists soon followed. Cruise ships travel to Antarctica during the summer, depositing people on the white continent, where they climb on ice cliffs and pose for pictures in their puffy parkas, standing next to penguins.

Today it is penguins, tomorrow it will be the vacation package to Mars.

MARS COLONY

Robert Zubrin's plan to go to Mars and back is based on the nineteenth-century example of Norway's Roald Amundsen, the first man to reach the South Pole and the first to sail the Northwest Passage. It is interesting to note that the Amundsen expedition was the smallest ever to attempt the Passage.

Like Amundsen, a manned Mars expedition could "live off the land" rather than carry everything it needed. It could even convert the carbon dioxide in the Martian atmosphere into liquid oxygen and methane to power the spacecraft home. This would enable a pair of 120-ton ships to go to Mars and back, unlike NASA's exorbitant proposal of a 1,000-ton space station.

When NASA scientists realized their massive Mars project was unaffordable, they studied the plans by Zubrin and his colleague

David Baker, both of whom were then engineers at Martin Marietta Astronautics, and concluded the plans were technically feasible. NASA gave $47,000 to Zubrin to build a prototype of the machine for converting carbon dioxide into rocket fuel.

According to the most optimistic plans, it would take six months to get to Mars, with at least one year spent exploring the surface, then six months back. Two years, at minimum.

Each member of the manned mission would have to be an explorer in the active, physical sense, as well as mentally able to cope with the unexpected. They would be cross-trained in many working disciplines, including basic medical knowledge, the sciences from geology to astrophysics, and technical know-how to fix everything from a solar panel to the plumbing on a space suit.

Some people have even seriously considered the optimal size of the group, pointing to fiction for clues. Michael Collins, a former astronaut and the author of *Mission to Mars*, suggests that the crew should be made up of married couples, citing the instability of a group of singles.

183. In mankind's efforts to settle Mars, there could be international projects set up in remote stations to serve the main Martian colony. Artist Ron Miller presents his concept of a possible joint American-Russian outpost to search for water and rare ore samples.
(NASA/Courtesy of Ron Miller)

MARS TOMORROW: THE DAWN OF A NEW MILLENNIUM

184. After a few years, the first Martian colony will become the first Martian city. Here we see planetary settlers exploring and prospecting with a Mars lander (in the foreground), and, in the background, more settlers arriving and departing from the Martian spaceport.

(NASA/Courtesy of Robert McCall)

· ·

The explorers would live in habitation modules constructed on Mars. According to Zubrin's plan, the habitation module would be twenty-seven feet in diameter and have two stories. The top floor would contain bedrooms, exercise area, laboratory, and galley; the bottom floor would be used for storage, particularly for the rover that would be used to venture beyond the landing site. Zubrin even suggests that the developer could persuade research teams to sign on for a certain cost, thereby expanding the outpost to a "settlement."

Throwing open the doors to Mars could certainly revive the spirit of exploration in society. Overcoming obstacles has proved to be one of the driving forces of invention and always generates access to vast new resources. As Zubrin says, "Without a new frontier, our civilization is doomed to stagnation."

Consider each detail for the benefits that could be gained by mankind here on Earth. Recycling would play a crucial role on Mars. Nothing could be excluded or thrown away simply because it costs too much to bring it there. Development plans for the ultimate in recycling processes are needed so the explorers could reclaim and use virtually every molecule taken to Mars. Similar technology on Earth could solve the growing garbage disposal crises before it is too late to save our oceans and atmosphere.

As for agriculture, the best mix of chemicals and fertilizers to start operating farming facilities must be determined. And the explorer colonists would need to construct greenhouses that could capture the maximum of the beneficial sunlight while cutting out the harmful rays. With the destruction of our own ozone layer, this could be lifesaving technology for millions of people.

Power and fuel are other big questions. Again, more advanced technology is needed in harnessing solar energy. Panels must be developed that are miniaturized for maximum power output. Well, after all, it was not so long ago that the microprocessor was thought to be impossible. Why not solar micropanels? It would lower the cost of transport, yet provide a never-ending source of fuel.

Unfortunately, society is fond of the already established precedence of nuclear reactors to generate cheap, powerful, and almost limitless energy. Most of the designs for Mars exploration and settlement involve nuclear power, which is why they are so expensive.

The one thing a manned mission has going for it is the indication that there is water on Mars. Scientists believe that water may still be below the surface in large repositories. These repositories could be mapped by robotic missions such as the 1998 Global Surveyor, providing indications of the most likely places where water and life can exist. Luckily, reclamation is a technology we already possess; people have been reclaiming water from frozen tundra for centuries.

A manned mission to Mars is not unfeasible, but a successful, low-cost expedition is impossible unless technology can be found to surmount these basic problems.

TERRAFORMING MARS

Most scientists (and the best sci-fi writers) accept the fact that in order to colonize Mars on a long-term basis, the current conditions must be changed. What with Viking's discovery of subsurface water and cloud vapor, the promising notion of terraforming has risen from doubtful to a nearly realistic possibility.

The problem with colonizing Mars is altering the atmosphere so people do not have to wear pressurized suits. Mars has rich resources of carbon dioxide, nitrogen, and metals, yet it also has unfiltered ionizing and ultraviolet radiation streaming in directly from the Sun. An artifically created atmosphere would give Mars the proper pressure, temperature, and oxygen volume humans need (currently the atmosphere is 76 percent nitrogen, 16 percent carbon dioxide, and 8 percent argon).

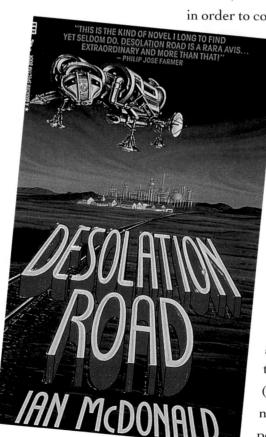

185. Ian McDonald's fantasy of Mars in *Desolation Road* (1988) maintains an ultra-romantic view while managing to cynically consider the reality of terraforming the harsh Martian ecology. The tone is reminiscent of magic realism, yet also reflects echoes of Ray Bradbury's *Martian Chronicles*.

(From *Desolation Road* [jacket cover] by Ian McDonald. Used by permission of Bantam Books, a division of Bantam Doubleday Dell Publishing Group, Inc.)

The Martian ambient temperature is always below freezing, despite the fact that the surface temperature can rise as high as 70 degrees Fahrenheit. But the Martian atmosphere is so thin it cannot retain the heat absorbed by the surface, and it is released into space. In effect, what Mars has is temperature without heat. The few molecules (as compared to the wealth of Earth's atmosphere) whip around with furious speed. There are not enough of them to transfer the kinetic energy of their speed into heat when they strike a surface such as skin.

Carl Sagan was one of the first to scientifically tackle the idea of terraforming in *The Cosmic Connection* (1973), while Gregory

186. Terraforming Mars looks like a much more difficult prospect than the fiction writers make it seem. In this image the only water on the surface forms as a thin film of frost on crater walls. Yet even reaching our neighboring planet has involved dozens of technological advances and leaps of imagination. Perhaps humans will eventually transform Mars into a world where we can live and prosper. (NASA/JPL)

. . . .

187. This NASA photograph of a full Earth dramatically reveals that the necessity of a life-filled planet is water. Earth is 78 percent liquid. (NASA)

Bear actually suggested that Mars should be moved in order to make it habitable for humans in *Moving Mars* (1993).

But the pinnacle of science fiction terraforming on Mars can be found in Kim Stanley Robinson's Mars series: *Red Mars* (1992), *Green Mars* (1993), and *Blue Mars* (1996). The trilogy dealt with the hardships endured by colonists as they tried to transform the barren wasteland into a place fit for human habitation. Yet throughout the series there was a counter-movement of people who insisted that Mars should be left in its pristine state.

Robinson described everything in an ultrarealistic style, from engineering a large underground habitat to genetically manipulating plants that would flourish under Martian conditions. He also suggested that ice meteorites could be used to bombard the world, raising dust and providing additional water. And, even more dramatic, he described boring deep tunnels through the crust to bring heat to the surface from the molten core.

Perhaps some of these techniques are not as far-fetched as fiction. Zubrin worked with Chris McKay, a planetary scientist at NASA's Ames Research Center, to come up with a plan for terraforming Mars to make it fit for long-term colonization. They propose to release chlorofluorocarbons (CFCs) into the Martian atmosphere.

CFCs are doing deadly damage to our ozone, but it is exactly what Mars needs. With enough CFCs released, an accelerated greenhouse effect could be started, warming the planet, melting the polar caps and subterranean frost, thickening the atmosphere with carbon dioxide released from the ice and soil. As plants are grown, oxygen would be produced, and eventually the air would become breathable.

Atmospheric thickening may also be possible through artificial surface impacts, similar to the natural effect of asteroid collisions and volcanic eruptions on Earth. Dust helps to contain heat and allows moisture to exist without immediately being sublimated to ice. Since the rotation speed of Mars is so similar to Earth's, terrestrial weather would be created in the form of similar wind and cloud formation.

188. Spring on Mars is captured in this Hubble Space Telescope view of the plant, the clearest single image ever taken from Earth. The question lingers on—will we find life on Mars?

(Phillip James, University of Toledo/Steven Lee, University of Colorado, Boulder/NASA)

. . . .

Only one thing is certain when it comes to Mars: no matter how we reach our neighboring planet, we will always dream of a future and an Earth-like Mars. We will always look for Martians, even if we have to become Martians ourselves.